Symphony of Matter and Mind

Part Five

TECHNOLOGIES OF THE MIND

The Brain as a High-Tech Device

Stanislav Tregub

TREGUB S.V.

Copyright © 2021 Stanislav V. Tregub

All rights reserved

No part of this book may be reproduced, or stored in a retrieval system, or transmitted in any form or by any means, electronic, mechanical, photocopying, recording, or otherwise, without express written permission of the author and publisher, except for short citations in relevant context.

For information about permission to reproduce selections from this book, please, write to symphony@stanislavtregub.com

Symphony of Matter and Mind. Part Five. Technologies of the Mind. The Brain as a High-Tech Device.

by Stanislav Tregub

ISBN 9798452552406

Cover design by: Ekaterina Volkova

The author is not responsible for the websites to which there are links in the book, and does not guarantee that the content of these sites will remain intact and relevant to the topic.

To my daughter

TABLE OF CONTENTS

Introduction vii
Chapter 1 Solving the Hard Problem of Consciousness 1
Chapter 2 Qualia Creation Technology 8
Chapter 3 Signal Identification Technology 29
Chapter 4 Technology of Overcoming Physical Limitations 49
Chapter 5 Memory Technology 64
Chapter 6 Neural Code Transmission Technology 90
Chapter 7 Movement Control Technology 115
Chapter 8 The Universal Processor 145
Annexes 155
References 167

INTRODUCTION

The brain is the source of sensations, emotions, desires, thoughts, memories, movement and behavior control. All these are aspects of the process we call the Mind. Despite a vast amount of data on the nervous system functioning down to the molecular level, no concept has yet uncovered the physical mechanism and the technology of this process.

With this aim in sight, the author continues to develop the Teleological Transduction Theory. The book contains hypotheses about the physical nature of the Mind and provides examples of how physics manifests in the nervous system physiology. It also shows how the Mind's algorithm produces a reality model with constant updating based on incoming data and performs the self-learning functions.

The theory encompasses the physical processes that create the enormous capacity, speed and multi-level complexity of our memory. It solves the riddle of how the brain forms and reproduces a vast number of representations almost instantly.

Building a model of reality is not an end to itself. The final goal is to act based on this model. The nervous system specializes in controlling the body and organizing purposeful movement. But how does it perform the function? The book contains hypotheses about the technology and physical mechanism that create the observed speed and efficiency of motion control.

Taking all these aspects together, the proposed theory aims to cover the explanatory gap about the physical nature of the Mind.

Chapter 1

Solving the Hard Problem of Consciousness

No theory can ever explain why anything is — that is the supreme mystery. But theory may be able to tell us why one thing rather than another is created and experienced.

Julian Barbour

Here is how the modern philosopher David Chalmers postulated the hard problem of consciousness: "It is undeniable that some organisms are subjects of experience. But the question of how it is that these systems are subjects of experience is perplexing. Why is it that when our cognitive systems engage in visual and auditory information-processing, we have visual or auditory experience: the quality of deep blue, the sensation of middle C? How can we explain why there is something it is like to entertain a mental image, or to experience an emotion? It is widely agreed that experience arises from a physical basis, but we have no good explanation of why and how it so arises. Why should physical processing give rise to a rich inner life at all? It seems objectively unreasonable that it should, and yet it does. If any problem qualifies as the problem of consciousness, it is this one. In this central sense of "consciousness," an organism is conscious if there is something it is like to be that organism, and a mental state is conscious if there is something it is like to be in that state" (Chalmers, 1995).

Chalmers is not the first and the last one to ask the fundamental ontological question about the Mind, and no one ever thought it to be an easy one. But his statement that the problem is hard produced such a 'wow effect' because it was pronounced at the right time and in the right place. The time has come to remind the scientists that while they tried to sweep the question under the carpet, it has not disappeared anywhere.

Postulating the "hard problem of consciousness," Chalmers asks why and how subjective experience arises. These are two separate questions. 'How' is a strict technological question. It is not an easy but a soluble task. We can say that if we cover technological issues sufficiently for understanding the process, we will answer the question. But 'why' is ambiguous. If we look at 'why' as a functional question, it contains the answer.

Question: Why, when our cognitive systems process signals coming through sight and hearing, do we have visual or auditory experience? Answer: Our cognitive systems create signals' representations which are the experience. Question: Why does physical processing give rise to a rich inner life? Answer: A rich inner life is the physical process of coding signals into representations. Question: Why does the brain produce consciousness? Answer: The function of the brain is to carry out the processes that we call consciousness.

Of course, there are many aspects to the process, and the brain is not an easy object of study. But conceptually, the question is not hard. It is even trivial if looked at from the functional side. That is what Chalmers calls the easy problem. His point is that 'why' is not a functional question.

Here is his argument: "Why should physical processing give rise to a rich inner life at all? ... What makes the hard problem hard and almost unique is that it goes beyond problems about the performance of functions ... Why is the performance of these functions accompanied by experience? A simple explanation of the functions leaves this question open ... Why is it that when electromagnetic waveforms impinge on a retina and are discriminated and categorized by a visual system, this discrimination and categorization is experienced as a sensation of vivid red? We know that conscious experience does arise when these functions are performed, but the very fact that it arises is the central mystery ... To explain experience, we need a new approach" (Ibid).

Chalmers is asking why consciousness happens at all. This is the problem for him. Nothing we will say about how the brain produces subjective experience will satisfy the philosopher.

What does Chalmers suggest as a new approach? He calls his position "naturalistic dualism." The term sounds new but rings an old bell. He tries to convince us that the only way of solving the hard problem is to acknowledge that that Mind is some fundamental entity that "goes beyond what can be derived from physical theory, ... over and above the properties invoked by physics" (Ibid).

That does not sound new at all. It is the 'good' old dualism. But Chalmers says that "it is an innocent version of dualism, entirely compatible with the scientific view of the world ... Nothing in this approach contradicts anything in physical theory; we simply need to add further bridging principles to explain how experience arises from physical processes" (Ibid).

This really sounds like a hard problem: we should explain something that is beyond physics by physical principles. Explaining physical phenomena by non-physical entities is an old tradition and an easy way out. But the other way around is something new. That is why the philosopher insists that "the moral of all this is that you can't explain conscious experience on the cheap" (Ibid).

We should become dualists and monists, idealists and materialists, mystics and scientists simultaneously. It is an oxymoron that produces cognitive dissonance. The moral of all this is that the philosopher sets an impossible task. It is not solvable, cheaply or expensively. Here we should just agree with Chalmers that scientific methods "must fail" (Ibid). The reason is simple: explaining non-physical phantoms is not within the scope of science. Let them remain within the purview of philosophy and theology. The scientific community should go on with building a physical model of the Mind as a physical phenomenon. There is no dissonance in this position, only harmony.

In this context, another quote from the same article is more important: "The fundamental laws of nature are part of the basic furniture of the world, and physical theories are telling us that this basic furniture is remarkably simple. If a theory of consciousness also involves fundamental principles, then we should expect the same. The principles of simplicity, elegance, and even beauty that drive physicists' search for a fundamental theory will also apply to a theory of consciousness" (Ibid).

Theory of Energy Harmony (TEH), developed in the first two volumes of this study called "Symphony of Matter and Mind," is exactly such a fundamental theory that offers principles of simplicity and harmony in explaining the "basic furniture of the world" (see "Part One. Music of Matter," "Part Two. Theory of Energy Harmony"). The Teleological Transduction Theory (TTT) developed in the following parts involves these fundamental principles as the basis for a physical model of the Mind.

Any scientific explanation should be based on hypotheses about a physical mechanism that could be confirmed or refuted. If the model does not have this attribute of a scientific theory, it may be due to insufficient accuracy in defining the laws and mechanisms or the presence of unknown factors that determine the behavior of the phenomenon under study. In this case, it remains only to admit that there is no satisfactory scientific explanation yet.

According to another modern philosopher, Joseph Levine, this is precisely what happens in the study of the Mind: there is an explanatory gap because we lack an explanation of the mental in terms of the physical (Levine, 1983).

Suppose we say that the Mind is a non-physical thing that has settled temporarily in the physical body. In this case, we answer the 'what' question but not the 'how' question about the principles and mechanisms of operation. If we declare that the Mind is some "fundamental property" of the world, we create the illusion of an answer as it still misses the goal of explaining how it works. That is why there is no difference between classic dualism and modern "innocent dualism." Both versions are incompatible with the scientific view of the world, and both are cheap explanations.

The question of how our subjective experience arises is about how our brain produces verbal, visual, sound, gustatory, tactile, olfactory, proprioceptive, interoceptive, painful, motor, and other representations. But for a dualistic position, how the brain performs this function and how the experience arises are different things.

Moreover, the technological question 'how' is entirely obscured by a shade of indefinite 'why.' This leads to a lot of fog about the "central mystery," which is accompanied by sloppy handling of strict technical terms. It is fine if we want to philosophize about the transcendental and fundamental, but it is a dead-end for answering the practical questions.

For example, Chalmers equates signal processing with information processing. This category mistake is not as insignificant as it may seem. Strictly speaking, our brain processes signals and generates information as code patterns that represent these signals. The quality of deep blue is a representation of a particular spectrum of light reflected from an object and perceived by the visual system. The sensation of middle C is a representation of sound vibrations of a specific frequency.

From this technological perspective, it is not perplexing that "there is something it is like" to experience a mental image because a representation is an image and subjective experience. The question "why is it that when our cognitive systems engage in visual and auditory information-processing, we have visual or auditory experience" does not make sense because signal processing done by the brain produces information that is the experience. But if we think that information created by the brain and subjective experience are separate things (as Chalmers suggests), we can go on wandering about their causal relationship forever.

As one of the pioneers of artificial intelligence, engineer Marvin Minsky noted: "Now, a philosophical dualist might then complain: "You've described how hurting affects your mind — but you still can't express how hurting feels." This, I maintain, is a huge mistake — that attempt to reify "feeling" as an independent entity, with an essence that's indescribable … When a mental condition seems hard to describe, this could be because the subject simply is more complicated than you thought. The way to get unstuck is to describe architectures with more details. Only then can we imagine how certain situations or stimuli could lead a brain into the activities that we recognize when we feel love or fear, or pain" (Minsky, 1998).

But saying that the brain activity produces subjective experience is not enough. That has been said in one way or another for centuries. The lack of progress in explaining the physical mechanism has fed all kinds of dualisms. Classical dualists say that it is something non-physical. New 'naturalistic dualists' say that it is over and above or beyond the physical.

Suppose we agree with Chalmers "that experience arises from a physical basis, but we have no good explanation of why and how it so arises." Do we need to dive into deep but essentially cheap dualistic metaphysics, as he suggests? No, we just have to find a good, physically substantiated explanation. And this is not a cheap way out. We have to show what the representations are physically and what mechanism creates this subjective experience. The question "What is the Mind?" implicitly contains the question "How does it work?" And it is not an idle philosophical issue but a practical task: we have to know the workings of the Mind to be able to fix its pathologies.

Let's see what encyclopedias and textbooks have to say about representations. The current Wikipedia article states: "A mental representation (cognitive representation), in philosophy of mind, cognitive psychology, neuroscience and cognitive science, is a hypothetical internal cognitive symbol that represents external reality or else a mental process that makes use of such a symbol ... Mental representations (or mental imagery) enable representing things that have never been experienced as well as things that do not exist. Think of yourself traveling to a place you have never visited before, or having a third arm ... Mental representations also allow people to experience things right in front of them — though the process of how the brain interprets the representational content is debated" (Wikipedia "Mental representation").

The word representation comes from the Latin repraesetare, i.e., symbolize. If we put this meaning in the above definition, it turns out like this: a mental symbol is an internal cognitive symbol that symbolizes external reality or a mental process that creates such a symbol. This is a useless tautology: a symbol is a symbol as a symbol. The author tries to explain something, but since the initial definition is about nothing, the subsequent explanation resembles treading water.

Are they images of something that does not exist or a representation of external reality? Is it something inaccessible to the senses or never experienced before? Is it a representation of the non-existent or something in front of you? As a result, the author draws a sad conclusion: science fails to represent how the brain represents representational representations. It is not surprising that many modern researchers negate the very concept of representation. If there is none, there are no problems with explaining how the brain creates the content of the Mind.

But representations can indeed be as diverse as the article describes. They can be images of non-existent and sensory experiences of real things. They can be a model of reality, or they can be phantoms. Dreams and imaginations, hallucinations and delusional constructions are also representations but not of current signals. The problem is not that the article's author has mixed everything into a "heap," but that there is no conceptual solution to the main question: how does the brain create this "heap"?

The author of the Wikipedia article at least tries to give some definition and description. In neuroscience textbooks, this topic is ignored. The word itself is used all the time, but there is no hint of what it means. As if everything is clear anyway. Representation is representation; why explain?

For example, the textbook "Neuroscience" (Purves et al., 2012) is full of such phrases: representation of the visual field, representation of movements, representation of the body and its current interaction with the environment, cortical representation, distributed representation, central representation, representation of a visual image by neurons, representation of sound frequency, topographic representation of the sound field, representation of different types of complex sounds, representation of natural stimuli, representation of olfactory information, representation in taste buds, orderly representation of information, symbolic representation, representation of words (language), neural representations, sensory representation, internal representation.

But you will not find a single word about what representations are physically and the technology of their creation. There are vast layers of anatomy and physiology, but there is not a word about what physical mechanisms are implemented in the described biological substrate.

Moreover, despite the constant use of the word "representation," some authors of the textbook abandon this concept in their articles. For example, Dale Purves wrote: "It would be best to describe visual perceptions in terms similar to those used to describe pain, for which the concept of representation makes no sense. Visual perceptions, like the perception of pain, do not stand for the properties of objects in the physical world, although the world, of course, generates the relevant stimuli. The distinction between visual perception conceived in terms of the awareness of behaviorally useful qualities vs. conceiving perception in terms of representations of the physical world would be a philosophical point only, were it not for the associated neurobiological implications. If vision does not represent the properties of objects and conditions in the world, then neither do its underlying anatomical and physiological mechanisms, which must, therefore, be thought of, examined, and tested in different terms" (Purves et al., 2010).

The author wants to say that the Mind is not a mirror image of the world but a modeling process. But lack of a clear technological approach makes it sound almost absurd. If vision is not about representing objects' properties and world conditions, what is it about?

These contradictions stem from the fact that the concept of "representation" has not yet received a physical definition and remains at the general philosophical level. But there are "implications." More precisely, there is the same question: what are representations physically, and how are they produced technologically? If representations are the product of the Mind, then we need to define it in strictly physical and technological terms too. Here it makes sense to repeat the basic hypothesis within TTT.

Hypothesis:

The Mind is the process of transducing signals from the external environment and the body into the internal code patterns representing these signals and constituting a model of reality for the purpose of active adaptation to this reality and maintaining the integrity of a living system.

The process includes the following stages:

1. Discretization and quantization of incoming continuous signals (analog-to-digital conversion).

2. Amplitude, frequency and phase modulation of signals.

3. Creation of wave patterns that represent the encoded signals (digital-to-analog conversion).

4. Integration of representations into a unified and differentiated model of reality by synchronizing the oscillatory activity of the filters that produce these wave patterns.

5. Saving representations as settings of the filters' impulse responses.

6. Projection of representations and comparison with the introjected signals.

7. Correction of the model in case of discrepancies with reality.

8. Evaluation, modulation and control of state of the system and its external actions based on the reality model.

These stages form an operationally closed loop of an iterative algorithm of Perception-Apperception-Action Lemniscate (PAAL):

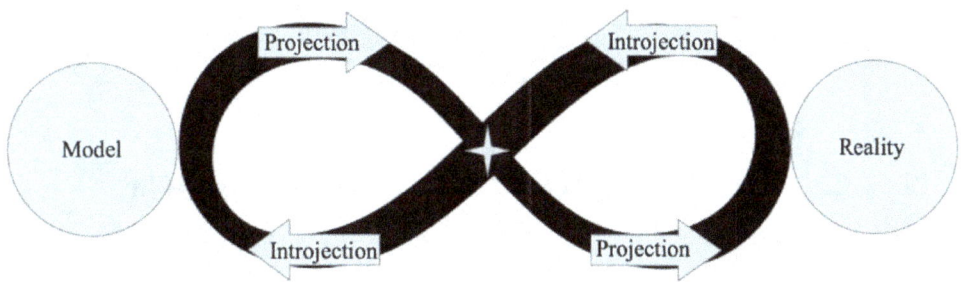

Thus, representations of the external and internal environment signals are wave structures of the neural network activity in the process of analog-discrete and discrete-analog transformations in the PAAL algorithm. This is what we call thoughts, images, sensations, feelings, etc. They have the same physical and technological origin. The difference between various levels of representations is in specific code patterns and network elements involved in forming this pattern.

This approach does not make any fundamental distinction between the representations of any signals from the internal and external environment, be they visual, auditory, painful, or other stimuli. The brain exists to create representations of all possible signals for adaptation and orientation in the external environment and control processes in the organism's internal environment.

Such definition speaks of specific physical processes in a material medium, i.e., it does not close in a tautological circle but leads to disclosing the mechanisms of these processes. On the one hand, it simplifies the task by providing clear guidelines for further movement. But, on the other hand, it significantly complicates it in comparison with the usual definitions at the level of "philosophical points," as it leads to the need to reveal related "implications." Without explaining the physics and technology of the process of creating representations, it is impossible to build a full-fledged "bridge" over which we can "walk" from the manifestations of the process to their explanation and back. The Teleological Transduction Theory in general and this part of the study in particular are devoted to describing the details of the process.

CHAPTER 2

QUALIA CREATION TECHNOLOGY

We invest stimuli with meaning, and apart from such investment, they are informationally barren.

Fred Dretske

The term 'technology' is usually attributed to machinery and tools. But its general application includes all human activities in interaction with the environment. Without any exaggeration, we can expand this concept to the activities of all living systems because any organism interacts with the environment to achieve the desired result.

The goal of a living organism is life. We can expand the description of this goal so that it does not sound as a tautology. Staying alive means that an organism keeps its internal processes functioning and remains an integral system that is separate from the environment but interacts with it and adapts to it. This is the general function of a living system as a whole. The concept of function (from Lat. functio — accomplishment, fulfillment) is used here as an activity with a purpose.

The living systems range from unicellular to multicellular with many subsystems. But even the simplest ones have many components that have a functional specialization. It means that each subsystem has its own teleology (from Old Greek τέλος — goal) within the overall purposeful activity of the system. So, the term 'technology' can be used not only for external actions but for internal processes which are also technological in the sense that they use certain tools and methods to achieve specific goals. The object of this study is the brain as a subsystem of a multicellular organism. Can we define its general function from a technological perspective?

The word technology consists of two words: techno (from Old Greek τέχνη — skill) and logos (λόγος — meaning, knowledge). Thus, we can interpret the term as a skill to create meaning and achieve knowledge. It sounds like a simple and clear description of the brain's general function: creating meanings from the signals of the environment and achieving knowledge about the world. Using technical terms, we can say that the brain transduces the signals into representations as internal code patterns that carry meaning within the semantics of this code. The teleology of this signal transduction process is to form a reality model as the tool for adaptation to the environment and management of purposeful activity of the whole organism. This is the essence of the definition of the Mind within the Teleological Transduction Theory (TTT).

So, the brain is a technological device, and the Mind (Consciousness, Psyche, Soul) is a technological process. This conclusion is not at all trivial for two reasons. First, it contradicts the millennia-old belief in the Soul is an immaterial entity temporarily residing in a material Body. This myth is the result of the objectification error when a process going on in an object was thought to be an object within another object. There are many reasons for this error but here we will stress just one: lack of knowledge about the actual physical processes within our body.

Second, the hypothesis sounds simple, but it has complications. Proposing that the Mind is a physical process going on in a physical substrate leads to the necessity of explaining how it works. It is a theorem that needs to be proved. If we leave it as an axiom that does not require proof, this approach will not differ from the belief in the immateriality of the Soul. Such faith is also based on an axiomatic statement that cannot be proved right or wrong. Any dogma that requires a belief means a lack of knowledge. The concept of the Mind that is developed within TTT is not a system of beliefs but a scientific model that is aimed at achieving knowledge. It requires the skill of creating meanings. We can say that it is a technological concept produced by the Mind to explain its own technology.

How can we make empirically grounded assumptions about the technologies of the Mind? As we have proposed in the previous parts of the study, it makes sense to proceed from physical analogies with processes that have similar functions. The technology may be more or less complicated, but this is just a difference in level. Simple technology can perform a similar function, and methods can be similar, but with a smaller range than a complex technology. This helps to build analogous models of complex and unclear systems using known and less complex ones. But such a reduction must be treated with caution as the analogy does not mean identity. It is essential to maintain a critical assessment of the model without equating the analogous system and the target system. An equalization error can make a useful analogy counterproductive.

Keeping this danger in mind, we will again return to artificial signal-processing solutions to illustrate the technologies of the brain. Let's take an example of holography. Here is a brief definition: "Holography is a technique that enables a wavefront to be recorded and later re-constructed" (Wikipedia "Holography"). Now, doesn't that sound like a function of the brain that records the waves of the incoming signals and reconstructs them as representations of these signals? If the technological aspect is the same, then, perhaps, there will be coincidences in the physical mechanisms.

The idea that brain processes can be analogous to holographic principles is not new. For example, there is the Holonomic Brain Theory (Pribram, 1991). However, this theory offers no explanation of how the brain creates meanings, which is its main function. Though it is called a theory of the brain, it contains no model of the neural encoding algorithms and their physical implementation. It is not enough to find a good analogy. For a full-fledged analogical model, it is necessary to explain the operation of the target system using an analogous system. Otherwise, the analogy turns into a general metaphor without specific content.

The holographic method of processing light signals was created in 1947 by Denes Gabor (Nobel Prize 1971). Gabor initially worked on the very narrow problem of increasing the resolution of an electron microscope. During this process, he incidentally invented a new method of photography and gave it the name "holography," which means "complete recording" in Greek (Gabor, 1948).

Regular photography only registers the intensity (amplitude) of light waves reflected from an object. This is enough to create an acceptable image on a 2D medium, but one parameter is not enough for full representation. Such a flat picture produces volume due to "circumstantial evidence": perspectives, the play of light and shade, occlusion, smearing of distant objects. But even stereo photography, which can create volume through the use of the properties of binocular vision, does not allow inspecting an object from different angles. Something is missing in such a representation of the object.

Light waves, like all others, have not only amplitude-frequency characteristics but also a phase portrait. To assess its dynamics, a frame of reference is required. It must also be a wave to "talk" with the object wave in the same language. Gabor made a smart move: he added reference waves ("coherent background") to the shooting process to register phase changes in object waves. Everything ingenious is simple if you formulate the essence: a phase, as development in time, requires reference points, a precise measure structure. The interaction of the reference and object wave structures creates an interference pattern fixed on the photosensitive element.

The carrier saves information about the parameters, but the picture does not look like a usual photograph. It is not a mirror image but a representation, as a particular pattern that stores information about the original signal. At first glance, the code created during holographic shooting and recorded on the carrier seems to be a random accumulation of stripes, specks, spots. But when a reference wave is projected onto it, a "miracle" occurs: the parameters of the original object wave appear, and an image is obtained as a wave representation of the signal introjected by the system. A voluminous image is created with all the accompanying details and depth effects. It looks like a miracle: we can view an object with varying degrees of approximation and from different angles, while its representation on a medium is a set of discrete patterns that are not a "mirror image" of the signal.

But our brain performs such a miracle all the time: it creates what we experience as a diverse world from a set of patterns stored on the carriers of the neural network. There are no landscapes, scents and sounds in the brain, but they are created and recreated there. Such a technology for constructing representations of environmental signals, which makes it possible to fit a vast external universe in a small internal universe, is the result of billions of years of evolution.

Artificial holography is still very far behind. There are reasons for this. Initially, Gabor had a problem: to create a truly accurate representation, the reference wave must be highly coherent and stable over the recording time window. There were no technologies for creating sources of such a wave at that time. He used a conventional mercury lamp, and the picture quality was insufficient because the reference structure itself was not clear enough.

But nothing prevented from doing the same in other areas, where the requirements for accuracy were not so high. For example, in radar, the holographic principle was used already in the 1950s. The antenna's coherent wave made it possible to take aerial photographs in poor visibility conditions. The best times for an optical hologram came in the 1960s, when it became possible to use a high-coherence laser wave. By the way, scientists who took part in the secret military development of radar holography were later also participants in the first experiments to create an optical hologram using a laser (Leith, Upatnieks, 1964).

But creating a stable reference wave didn't fix all the problems. There is one more requirement for the efficient operation of the technology, which is still the most serious stumbling block for the full development of the principle of holography in artificial systems. To provide an efficient and reliable memory, a suitable material is needed as a medium capable of storing all complex interference patterns. Artificial technologies follow the path of developing magnetic, microelectronic and optical carriers, which do not yet provide such an opportunity. Living systems have solved this problem with sufficient efficiency in another way (more on that in the following chapters).

So, the essence of creating representations is to combine the object wave and the reference wave, and then the reference wave and the result of the previous combination of two counter flows stored on the carrier in an encoded form. Initially, the wave interference pattern is encoded as discrete patterns, as states of the substrate elements, and then reproduced as a wave structure. If you connect the chain into a constant iterative algorithm, you get the PAAL: the projection of representations is superimposed on the current introjection, the interference pattern is saved as new representations, and again according to the algorithm. Requirements for the reference wave: stability and coherence. Requirements for the interaction of the reference and object waves: synchronization as a stable frequency and phase coupling (more in that in "Part Six. Harmonies of the Mind").

But how does the process of superposition of the object and reference waves take place in the process of the Mind? Maybe our brain emits certain waves towards the signals of the environment? It is appropriate here to recall the history of ideas about the work of vision. The ancient Greek philosophers disagreed about the mechanism. Plato believed that the eyes send something outward and feel the world (extramission theory). Democritus believed that the eye catches particles of a particular kind sent by all objects (intromission theory). Democritus's model was an analogy with digestion and Plato's model with touch. As is often the case, both were right and wrong at the same time. Consciousness is a process where there is both projection and introjection, but the physical mechanism, to put it mildly, is somewhat different than they imagined.

The concept of extramission (emission) has prevailed for a long time. The logic of the rationale was "proof by contradiction," i.e., it refuted the opposite concept. There are several arguments. First, if we absorbed everything we look at, we would simply blow apart. If the process of intromission involves the capture of matter, then a way of filtering and removing it is needed. It is necessary to digest the essential and throw out the excess. Intromission theory did not explain this mechanism. The question arose of how the same object enters the eyes of different people. Furthermore, why does it not disappear from such constant consumption? But the extramission model also did not explain the mechanism of the proposed "sight touching" of objects. Plato and Euclid wrote about certain rays emitted by the eyes, but the hypothesis remained pure speculation since no one observed these rays. By denying the opposite model, the other did not prove itself in any way.

Many centuries have passed since the time of Plato. In the eleventh century Hasan Ibn al Haytham (Alhazen), in his work "Book of Optics," showed the inconsistency of the primitive concept of extramission. His theory was related to the idea of intramission, but at a new level. He believed that natural light and colored rays affect the eye, and the image is obtained using the rays emitted by visible bodies. That is why, he wrote, the stars can only be seen in the dark.

Leonardo Da Vinci wondered: where is the origin of vision — in a visible object or the seeing eye? He hesitated between the concept of intromission and extramission. If the source is inside, how can the eye radiate itself? How can such rays reach the Sun and the stars? Da Vinci understood the absurdity of such an idea. Many phenomena confirmed that signals come from outside, but the Mind turns out to be a kind of "drain" for infinity. Is it a source or a sink? He could not connect both flows (outside and inside) in a physically plausible theory. He was left at a loss: how can such a tight space accommodate the universe of signals?

Here is how he describes the introjection: "Objects opposing the eyes act with the rays of their images like many shooters who want to fire from a gun into a certain hole. From these shooters, the one who turns out to be on a straight line between the hole and the gun will succeed. Likewise, objects opposing the eye will have the greater access to feeling the more they are on the line going to the hollow nerve. The water that is in the pupil, around the black center of the eye, acts just like a sniffer dog on a hunt. They find the beast, and the greyhounds grab it. In the same way, this water — it is moisture, which has something in common with the perceiving ability — this water sees many objects but does not grab them; however, immediately the ball in the middle turns there; it is located on a straight line that leads to feeling, and it captures images and imprisons those of them that please him in the prison of memory. The eye is the most wonderful thing in the world" (Shevchenko, 2013).

Indeed, the eye is a technological achievement of evolution, but no less wonderful "thing in the world" is the brain, which processes the signals that enter the eyes. Let's continue the metaphor. Sniffers are not enough: it is not enough to find a beast; you need to grab it, process it, and use it to your advantage. The questions remain. How the "greyhound" of the brain grabs the "beast"? What and how the other participants in the "hunt" do? What happens on this "straight line to feeling"? And is it straight? How are the images that fall into the "prison of memory" created? And how are they extracted from it? How does the Mind know what kind of "beast" is in front of it?

The introjection part of the process is interesting and complex. Still, even if we reveal all the physiological details of this "straight line to feeling," the feeling itself will remain a mystery to us, and we will be forced to return to the endless regress of the observer in the observer in the observer.

Two centuries later, Rene Descartes drew an introjection scheme based on more recent anatomical data:

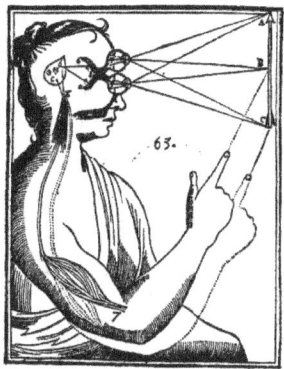

According to Descartes, the visual image is formed in the Pineal gland (Epiphysis), which he called the "seat of the Soul," the place where all thoughts, images, feelings and sensations live. The scheme is no longer a straight line, but a circular one: from objects to the eyes, then to the "place" of the Soul, from it to the motor effectors, and from them again to the objects. But this scheme remains linear: the algorithm of perception, apperception and action goes in one direction of introjection of signals and responses to them. Such a model existed in the cognitive sciences for almost 400 years, expressed itself in the standard signal-reception paradigm, and still exists today.

But although the path from outside signals to brain tissues inside has been studied to the intracellular level, the question remains: how does the brain embody consciousness? How does this "hunter" not just catch the "beast" but set a goal for himself on the hunt? How does it know what kind of animal there is, goes on its trail, finds, catches, processes? How is the complete algorithm implemented, not just searching and catching?

Descartes gave the following answer: "A picture forms and appears on the inner surface of the brain, opposing its concavity; from here I could transfer it to a certain small gland located in the middle of the concavity and which is, in essence, the receptacle of feelings" (Shevchenko, 2013).

But the question is, how does the picture form? Even if we find a thousand and one receptacles of feelings, the question remains, how are these feelings created in these places? If we answer the functional question, what do these places do, then the physical, physiological and technological question remains: how do they do it? The fact that Da Vinci, Descartes and many other scientists of past centuries could not answer this question is understandable: there was not enough empirical data. But with all the cosmic volume that we have accumulated to date we have no such excuse

In Descartes's scheme, as in all subsequent models up to the 21st century, at first explicitly, and then implicitly, a "homunculus" was required, as a special entity inside the brain that creates what we call the Soul. A "spectator" was required who simultaneously watches the external "cinema" of reality and creates an internal "cinema" of the Mind. Descartes painted a bearded spirit inside the head. No one draws such spirits in neuroscience textbooks, encyclopedias and scientific publications anymore. But without answering the question "How?" in all complex descriptions, diagrams, formulas and models, there is implicitly a "spirit" that does its job in a mysterious way.

Walter Freeman III wrote: "The loss of the Cartesian pilot has left a large gap in the theory, because no one wants a homunculus, but no one has a replacement" (Freeman, 2008).

We remain at the level of the dilemma described by Aristotle: "Since we perceive what we see and hear, it is necessary to perceive either by sight, what it sees, or by another feeling. If by sight, then it must perceive both sight itself and its object — color. So, either there will be two senses for the perception of the same, or the vision will perceive itself ... If the feeling that perceives the vision was different, and not the vision itself, then either the series would go to infinity, or some feeling would perceive itself" (Shevchenko, 2013).

What way out of this impasse of endless regress and cyclical logic did Aristotle suggest? The active role of consciousness: "The mind in action is what it thinks." The presence of a certain experience that creates our current perception of the world is an evident phenomenon, and philosophers of all times and peoples tried to explain it.

Emmanuel Kant believed that he was making a "Copernican revolution" when he argued that our consciousness not only passively comprehends the world but is an active entity. He made a distinction: "the thing as it is" (ding an sich, often translated as "the thing in itself") and the world as it is given in experience. It was radically different from the objectivist paradigm of knowing the world as it is "in reality." But gradually, science moved from philosophical hypotheses to the practical study of processes, and the ancient concepts of extramission and intramission were combined at a different level of knowledge.

In the 19th century, Hermann von Helmholtz created a theory of visual perception in which projection played a leading role. He called it "unconscious assumptions" (von Helmholtz, 1867). Helmholtz studied optical illusions and found an explanation for them: the brain projects the assumptions formed by experience, and they dictate our perception of the environment. Many examples can be given, but one is sufficient. As Helmholtz himself wrote: "Every evening before our eyes the sun sets behind a fixed horizon, although we know perfectly well that the sun is stationary and the horizon is mobile" (Ibid).

The model of reality "the Sun is moving and the Earth is standing" has been formed for billions of years in the experience of all terrestrial beings. Human knowledge about another aspect of reality has arisen recently. In our earthly experience, we still proceed from the geocentric model in our daily processes and actions. Our sensory-motor perception and apperception say: the Earth is standing and the Sun is moving. It is not surprising that the real Copernican revolution (the heliocentric theory) was not immediately accepted. It took a lot of time and new experience gained with new tools for measuring environmental signals. But even now, only cosmonauts in orbit have a real sensory-motor experience about a different picture of the world, seeing the process literally from another point of view.

In studying the introjection part, great strides have been made. The path from eye receptors to signal processing centers in the cortex has been sufficiently explored to more or less clearly explain how introjection (intromission) occurs. It is also clear that the eye is not a projector. But how, then, is the visual model of reality projected? And the same goes for other modalities. Everything again comes down to the central question of millennia: what is the Mind (Soul), and how does it work? In many ways, we remain at the level of the Ancient Greeks: the projection has not been explained, although it is considered a generally recognized phenomenon.

Here we do not mean the narrow understanding of projection in psychodynamic theories as a defense mechanism by which the ego protects itself by denying the existence of some qualities in itself and ascribing them to others. In TTT, projection is understood as the physical mechanism which ensures the current reproduction of representations created based on processing signals from the environment in the past.

For example, the phrase "yellow room" means that the subject projects representations based on processing a specific range of electromagnetic waves, called "yellow" at the level of verbal representations. There is yellowness in the object in the sense that it reflects light in the range that we have called yellow. The same applies to sound: it exists by itself in the environment as vibrations of air or another medium, and we call the representations created as a result of the transformation of these vibrations by the brain as sound. Reproducing these representations as patterns of neural activity can be called projection.

There are signals from the outside, but the final representation, as a combination of the projection of the past and introjection of the present, is inside. We see the world not as it is "in reality," but as our brain created it. At the receptor level, the nervous system works technologically as a receiver, processing and transforming the object wave. But it also works as a projector inside itself: it projects the reality model as a reference wave, combining it with the object wave.

We are so accustomed to the fact that our brain works very quickly (normally) and the imposition of the projection of representations on the introjected signals occurs smoothly (normally) that it is not easy to accept such a simple fact: our whole picture of the world is determined by how these two waves work in the algorithm of the Mind, how well they are synchronized and combined.

Hypothesis:

Brain filters transform waves of energy (signals) coming from the environment into representations of these signals. This generates object waves superimposed on the projected reference waves generated by the integrating filters based on these filters' existing impulse response settings. As a result of the superposition of the patterns of wave activity of the projection and introjection phases of the PAAL, the interference of the reference and object waves occurs.

The interference pattern allows object and reference waves comparison. In case of discrepancies that go beyond specific threshold values, the representations are corrected and remain in the system as new settings of the impulse response of neural networks and are again projected onto the introjected object waves in a constant iteration of the PAAL algorithm. As a result, the reality model retains both stability and dynamic adaptability.

The requirement for maintaining normal superposition and comparison of reference and object waves is synchronization, as a stable frequency and phase coupling. For the reality model to be coherent, stable and adaptive, the introjected and projected waves must be matched. This requires the synchronization of all filters from primary receptors to higher integrators. Synchronization with the environment and internal synchronization is a condition for adaptation and survival.

In case of desynchronization, leading to the appearance of breaks in the algorithm, the representations encoded and stored in memory continue to be reproduced, but they are not coordinated with the introjection of the signals of the environment (object waves) and among themselves (decay of the reference wave coherence). The function of testing reality is disrupted and phantom representations detached from introjection arise. They can arise under the following conditions: sleep as a dissociation of the system from the current signals of the environment (we call such representations dreams); with sensory deprivation, in an altered state of consciousness during trance dissociation (we call it visions); when exposed to psychoactive substances, in various pathologies of the nervous system, trauma, strokes and any conditions that affect the interaction of the entire chain of brain filters (we call it hallucinations).

The etiology of states can be different; the essence is the same: violation of physics and technology of coordination of the projection of the reference waves of the reality model and the introjection of object waves of environmental signals (more on that in "Part Eight. Dissonances of the Mind").

Physically and technologically, representations can exist on their own without input signals. Therefore, nothing is surprising in the fact that in a particular state, a person can see images, hear voices and other sounds, feel smells and tastes that are not present in the environment at the moment. It seems to him that he looks at these images with his eyes, hears them with his ears, smells them with his nose, and so on. The truth is that when a chain of an algorithm is disrupted, senses have nothing to do with these representations. This is a pure projection of the internal product of consciousness exported to the outside in isolation from import to the inside.

Indeed, even in a normal state we only look with our eyes, but we see with the Mind's eye, that is, with the brain. But nothing prevents us from seeing and hearing with our brain, even when our eyes and ears are out of work. It can occur during dissociation in the form of sleep or trance and with pathological disruption of the technological chain resulting from congenital or acquired disorders. We will consider pathologies in further parts of the study. Now we are interested in the physics, physiology and technology of the normal process.

To form representations of signals of the world the brain has to solve two problems. Measurement of signal parameters is the so-called direct (forward) problem. There is also a reverse (inverse) problem: reconstructing a signal from the measurement. The Mind is not a mirror that reflects the world, but an active function that creates the model of the world.

Let's continue with the visual picture of the world as it is the most important for humans. Physiology confirms this fact: the major part of the cerebral cortex is engaged with vision. The brain faces many problems while creating meanings from the light wave signals. They, like all other environmental signals, are multidimensional and potentially infinite. Our eyes are not just finite and small, but the retina is a flat matrix. How does our brain solve the reverse problem of creating a 3D world from an image obtained by a 2D matrix of the eye?

Partially the problem is solved in the primary converters of the retina. The surface flatness is compensated by the spherical shape of the matrix (approximately 70% of the sphere). If it were, for example, a square or a rectangle, then many parameters of the objects would be elusive. But in any case, the retina form alone does not create the volume. What is the technological solution? Stereoscopic vision.

We are so used to the fact that we have two eyes that we take it for granted. But there is a technological reason for it. One eye is not enough as different objects and their movement can be reflected on the retina of one eye in the same way:

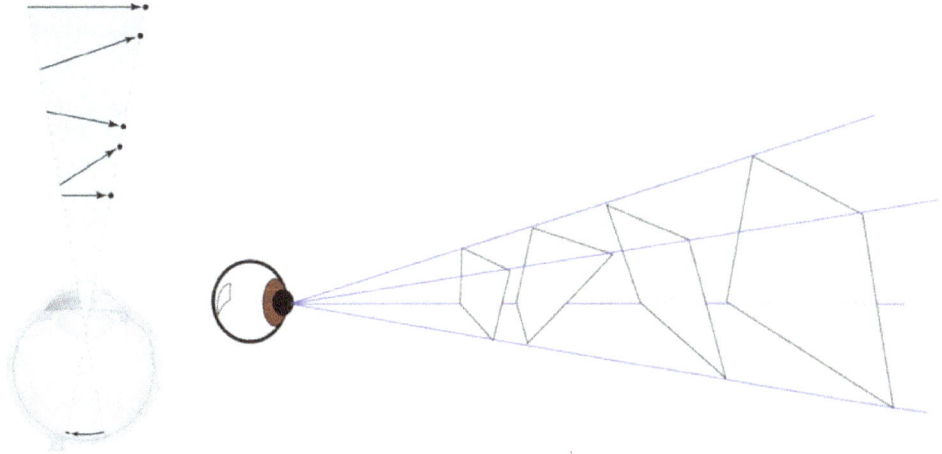

To define the shape and movement with sufficient precision we need two eyes. Some animals possess a third (parietal) eye. But it is rudimentary and, for the most part, plays the role of a photosensitive endocrine gland, participating in the regulation of circadian and seasonal rhythms. In modern animals, it is much less common than in fossils. The evolution of the analog part of the visual pathway has come to the optimal number of eyes to solve the reverse problem of reconstruction of light signals.

Stereoscopy creates the volume of the world. The image obtained by the retina of one eye differs from the image on the other retina due to the distance between them. By processing this difference, the brain creates a three-dimensional visual world. By the way, it is this property of brain technology that is used in 3D cinema, when two pictures are created during shooting from different cameras located at a certain distance from each other. Each viewer's eye sees only the part of a stereo pair intended for it (special glasses are used for this), and the brain perceives two images as one volumetric.

Stereoscopy can be called "direct evidence" of volume. But we also need "circumstantial evidence": shades of color, light-shadow transitions, occlusion, perspective, less sharpness of distant objects, parallax. People with two eyes, but with pathologies that cause a violation of the function of processing such signal parameters, perceive the world as flat (more on that in "Part Seven. Inner Universe").

How does the brain produce colors of the world? Again, this is a question of signal reconstruction technology. Objects do not have colors. This may sound strange as we are used to attributing colors to everything. But color is not the quality of the object. The quality of any surface is its ability to reflect a certain frequency range of light waves. Blue is not the quality of a blueberry, but the quality of the representation (qualia) that our brain creates from the reflected waves in the range of 600-670 THz. The same goes for all qualia of vision or any other modality: they represent the properties of the signals that our brain perceives and encodes.

The term "qualia" has acquired a mystical meaning thanks to some modern philosophers, but let's just stick with the simple one that stems from the etymology of the word "quality." It is about the question "what kind" (qualis). Clarence Lewis, who coined the term qualia, wrote: "There are recognizable qualitative characters of the given, which may be repeated in different experiences, and are thus a sort of universals; I call these "qualia." But although such qualia are universals, in the sense of being recognized from one to another experience, they must be distinguished from the properties of objects" (Lewis, 1929). In short, qualia are representations.

The color qualia of the world depend upon the ability of the system to receive and transduce a particular range. For a living system with the simplest proto-eye (light-sensitive cell), there are no colors in the world because it does not process these signal parameters. Some living systems do not see particular colors in the sense that their visual modality is not set up for the transduction of these frequency ranges.

Each additional parameter of a signal requires processing resources. But everything has its reason, at least in the technologies of the Mind. Each parameter indicates the state of external objects. A transition of the reflected light wave from one frequency range to another is a signal about the details of the object. And the more details, the more mature the berry we collect in the literal and figurative sense.

How does human vision create colors? Let's start with the primary conversion of the light signal. The matrix of our eyes has millions of receptors, but this value is incomparable with the volume of electromagnetic oscillations reaching the retina. Moreover, the receptors are very few in terms of the qualitative characteristics of the types. There are only two of them: cones and rods. The cones are divided into three types according to their sensitivity to different wavelengths: S-type cones are sensitive to violet-blue (S from short wavelength spectrum), M-type — to green-yellow (medium wavelength), and L-type — to the yellow-red (long wave). The division into three categories is formal since the areas of responsibility overlap each other, and these resonator-receivers can change their settings. Their impulse response has a specific range within their specialization, but it can change.

Any color can be expressed by the formula: $C = RR' + BB' + GG,'$ where RGB are the parameter values, R'G'B' is the relative weight (intensity). The retina does not have an equal number of cones for each range: approximately 65% red, 32% green, 3% blue. Red and green cones are in the center, blue cones are at the periphery and are most sensitive.

Rods also help: they are sensitive to some ranges (blue and emerald green), but they are mostly tuned to light-shadow perception. They are more sensitive and numerous (about 120 million compared to 6-7 million cones). They are important for the system: shades of colors are the necessary details, but it is the basic contrasts of light/shadow that create the volumetric world that we see. Cones lose their sensitivity when the signal strength drops, so we hardly distinguish colors at night.

All receptors are unevenly distributed over the retina. The system tries to solve the problem of a limited number of elements and, in general, the small space of its visual matrix. Technologically speaking, it distributes sensors over the perception field, which has a finite-support function. The rods are more focused on the periphery, which gives an advantage in the night and peripheral vision. There are almost none in the center. The cones cluster closer to the center as humans are predators and need to assess the state of the external target with forward-directed focused vision. But our matrix is not just a passive film that lies and waits for the signal to come running to the right place on its plane. The eye continually makes micromotions (saccades) to keep objects in the center of focus. The constant active work of the "camera" is needed to cover the huge external space with a small retinal space.

Using vision as an example, we briefly described how the brain solves the direct problem of measuring signals using the analog-to-digital part of the PAAL algorithm. For solving the problem of signal reconstruction, the brain performs the reverse operation of digital-to-analog conversion and projects representations. This way, the brain can compare the accumulated data with the current data.

In engineering, two separate diagrams of a direct and a reverse problem in information processing are drawn:

Direct problem: Object → Receiver → Measurement
Result: Object → Parameters of the object

Reverse problem: Object → Receiver → Measurement
Result: Measurement → Representation of the object

At first glance, the movement is going in one direction. But the direct approach answers the question: what will be the measurement of the object's parameters? The reverse answers the question: what will be the object's representation with these measurements? For a living system, solving the reverse problem is the ultimate goal of signal reconstruction. It needs to answer the question: what has been measured? The brain solves the forward and reverse problem of signal reconstruction thanks to the technological scheme of Perception-Apperception-Action Lemniscate (PAAL) with two-way flows:

1-2: Representation → Object
3-4: Object → Receiver → Measurement → Representation

Now let's see how these technological solutions are implemented in physiology. Here is the anatomy in a simplified form:

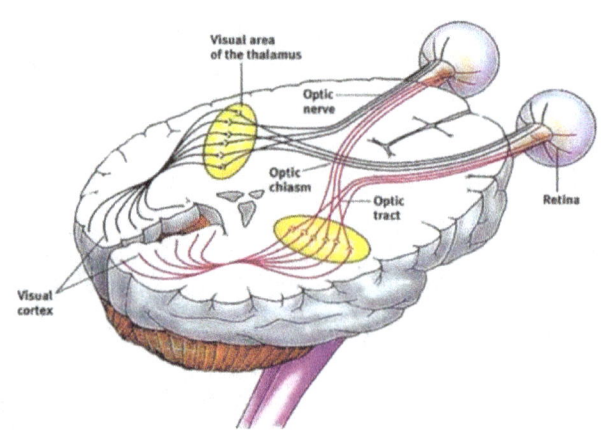

What happens to this physiological map when transformed into a functional diagram? Usually, they begin to connect all the elements of the chain into a linear circuit like a block diagram with boxes and arrows of input/output between them:

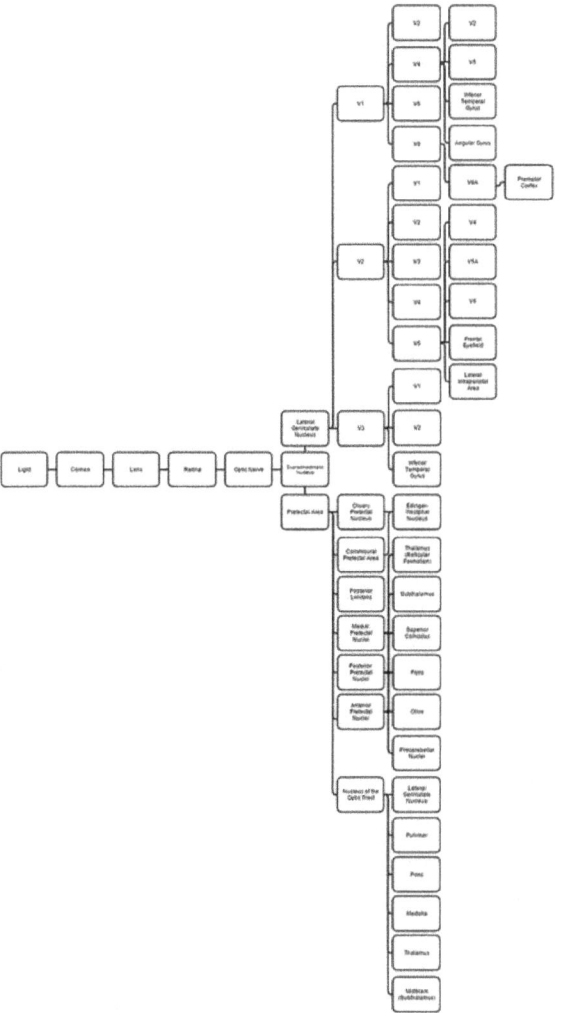

Wikipedia "Visual system"

There are much more complex schemes, but even this one does not fit on the page and has to be compressed. However, its most important aspect is clearly visible: linearity. There is a signal (box on the left), there is an entrance to the system (numerous boxes of the visual pathway) and there is an exit (box of the motor zone on the right).

It is the view on the brain as the receptor of signals. But let's suppose that the system does not work linearly in one direction, that it is a "two-way street." Such an assumption is banal in itself, but it requires an answer to a far from simple question: what happens in the system besides perception and action? How does apperception work? How does the brain create meanings to perform meaningful actions?

To begin with, let's draw a simpler and more intuitive functional and technological scheme. It will turn out to be strikingly close to the anatomical, where flows follow a pattern with intersection-decussation (optic chiasma) and feedforward-feedback connections.

Visual System Lemniscate

Processes:
1. The effect of light waves reflected from objects on receptors.
2. Signal processing — transduction, transmission, modulation, synchronization, creation of representations, storage, reproduction.

Perception mechanisms (primary processing and transformation of signals):
1. Photoreceptors (rods, cones).
2. Bipolar cells.
3. cGMP (Cyclic guanosine monophosphate) ion channels Ca^{++}, Na^+, K^+.
4. Conformational change: opsin/rhodopsin in cGMP.
5. Activation/deactivation of the ion channel.
6. Ganglion cells on-center and off-center.
7. Receptor fields — the spatial structure of receptor organization.
8. Horizontal cells as filters-transducers in receptor fields.

Lemniscate stages:
1. Deductive projection of the reality model. Visual cortical structures such as extrastriary visual cortex (zones V3, V4, V5, MT), parietal lobe (spatial analysis — Where?), inferior temporal lobe (identification and characteristics — What?), occipitotemporal gyrus (identification — Who?), and in the frontal cortex create and project a reality model.
2. Inductive projection: the reference waves of the reality model interact with the object waves of the introjected and encoded environment signals.
3. Inductive introjection of environmental signals. Specialized receptors perceive signals (light waves), transform them into an internal code, as patterns of neural activity, and transmit the created patterns to the corresponding thalamus module (lateral geniculate body), as a distribution relay and filter-modulator.
4. Abductive introjection. The information is further distributed among the layers and zones of the cortex for processing, integrating, evaluating, comparing,

saving, reproducing and correcting the states of the system and the reality model, transmission of signals to motor zones and effectors for action in accordance with the model.

Such a functional and technological scheme of the process works for all sensory systems with a difference in anatomical details (see "Annexes").

There is one more problem that the brain needs to solve when creating meaning. When light waves hit the retina, the image turns out to be inverted due to the optical refraction in the eye lens:

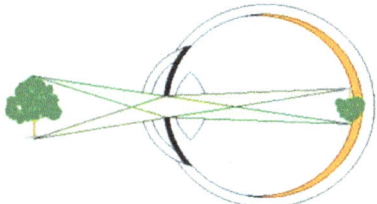

But we need to see the world not upside down, but in accordance with the position of our body relative to the gravitational base and the location of the parts of the body relative to each other. The vestibular apparatus and proprioception create a "grounded" picture of the world, and the physics of optical processes in the primary transducer (eye) turns everything "upside down": it creates a complete inversion horizontally and vertically. It is necessary to return everything to its place at the next stages of the technological chain.

The problem of an inverted image obtained on the retina as a result of optical signal conversion has been known for more than one century. But this is a problem for scientists to explain theoretically why we do not see the world upside down. The brain solves this problem practically and technologically. Throughout the history of science, there have been many versions of how it does it. And even very exotic from the modern point of view, but within the framework of the "good old" version of the Soul as a kind of entity sitting inside the body. They said that the Soul is simply upside down in the head, so everything is inverted back to its place. Not everyone was happy with this hypothesis, since some questions remained.

What is the up or the down of the Soul? How does this Soul see? This is again a version of the homunculus and screen in the head, leading to endless regress: the homunculus in the homunculus and the screen in the screen, and so on. And it still doesn't answer the questions. Why is the Soul inverted in the case of sight, and not inverted in the case of other modalities of perception? After all, we feel, for example, our heels as heels, and not as the crown of the head. We have an unstable Soul. Now its upside down and then down-up, or in different poses all the time. For those who are satisfied with the answer of all times and peoples about the transcendence and immateriality of the Soul, these questions are not interesting. For them, the Soul is what it is, how the Creator made it. Period.

But if we perceive the Soul as a physical process in a material living system, then this is a specific question: how does the Mind see? How does it create a picture of the world from the signals it receives from the retina? The first experiments, which clearly showed that it is not the picture on the retina, but what representation the brain creates that matters, was carried out at the end of the 19th century by George Stratton (Stratton 1896, 1897). He made upside-down goggles (an invertoscope) that flip the image that the retina of the eye gets and wrote a detailed report about his experience of wearing them.

He described it as the loss of harmony between vision and hearing, kinetic, tactile, vestibular senses. The representation of the visual field turned over predictably, and he began to see the world "upside down." But the rest of the modalities insisted on the usual way of world. For the brain, it was a sharp crisis of the gap between the general model of reality and the introjection of signals in one modality. It continued to project the familiar model, but the signals partially contradicted it. Some modalities continued to confirm what was developed by evolution under the conditions of gravity and our upright posture, but the eyes were knocked out of the general ensemble.

The system does not tolerate the schism of the model and signals. Can it change the signals? In this particular experiment a person can always take the invertoscope off thus changing the signals. But in general, we are not mastering the world, but adapting to it. So, imagine that you keep on wearing the invertoscope for the experiment to be consistent. What can the brain do, if the signals suddenly entirely contradict the model, that was worked out for billions of years of evolution? Can it change the model so fast? No. Moreover, the rest of the modalities are at odds with vision. Sounds like a technological dead-end: you cannot change neither the signals, nor the model. The trick is that you do not need to. This is a technological miracle, that the brain manages to do due to the primacy of the statistically verified reality model with high probability weight. The system is not a passive mirror of the world, but an active entity creating its inner Universe.

Let the retina filters-converters do their job and encode the new way the signals fall on it. Let the modulators do their job of transforming the encoded messages into parameters acceptable for further processing. But it is the integrators that produce final representations. They are the meaning creators. But if this higher part of the hybrid ADC-DAC chain of the brain works well it won't take long to get the picture back to normal, to reharmonize the senses. Even if you continue to wear the invertoscope your world will gradually get "down to the floor." The image falling on the retina will still be inverted relative to the usual inverted one. But, in any case, the brain will manage to stabilize it back to what it should be for adapting to the real conditions. The secret of the technological feat is the primacy of projection.

What conclusions did Stratton draw more than hundred years ago? "It is clear, from the foregoing narrative, that our total system of visual objects is a comparatively stable structure, not to be set aside or transformed by some few experiences which do not accord with its general plan of arrangement" (Ibid).

Exactly, this general plan is the projected reality model. Unfortunately, for most of the next century a "mirror" paradigm reigned in psychology according to which we are just stimulus-reaction machines. It's time to get back on the ground and invert our vision of ourselves. The brain is a highly technological mechanism that creates meanings from signals, not a black-box that just "swallows" signals. Higher integrators exist in order to unite together the indications of all perception modalities and eliminate the arising contradictions in favor of a version of the reality model that is adaptable to the given environmental conditions. They are able to turn the picture of the world in any direction, and the experiment with the invertoscope proves it. But in natural conditions it is not necessary to employ such a "software" solution. The integrators are high energy consuming elements. They should be busy with more important things than turning the world up-side down. There is a hardware solution to the inversion problem in the optical part of the chain.

David Hubel, winner of the Nobel Prize for the study of visual systems, wrote: "We do not know exactly why the retina is arranged in such a strange way — as if it is inverted and ... no one knows why the right half of the surrounding space is usually projected into the left hemisphere of the brain" (Hubel, 1988).

But let's just look at the physics of the process. The retina is not arranged in a strange inverted way. It is a two-dimensional matrix-screen, on which rays of light fall, which are inverted after refraction in the lens due to the physics of optical processes. To reverse the flip, the network topology diagram should be as follows:

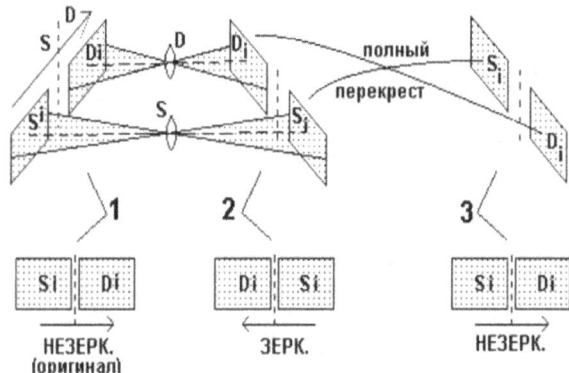

Voronkov, 2009

The diagram shows that to return to the original state of the light signal, you need to cross signal streams. Why not reverse the signal paths after the retina so that the physiological scheme coincides with the necessary movement of information? Perhaps this is a simple solution to the mystery of the intersection (decussation) of the visual pathways. This makes it easier to coordinate visual signals with signals from other modalities, especially vestibular and proprioceptive, which persistently speak about the position of the body relative to gravity (up/down) and the position of body parts relative to each other and external signals (up/down, right/left, front/back).

The flip can be done in different ways, and in evolution there was a selection of options by trial and error. For example, invertebrates have receptor axon decussation that is absent in vertebrates. Thus, immediately at the exit from the primary converters, a reversal of flows occurs. However, at the same time, differentiation suffers, since too early an intersection leads to the fact that information is largely processed at the level of simple elements of the system even before modulators and integrators. But they use what they have. Some invertebrates do not have higher integrators at all. But subtle differentiation is needed, and evolution again followed the path of increasing the sufficient and necessary complexity of structures.

The diversity of both the structure of the eyes and the ways of further transmission of information shows how the development proceeded. Genetic data indicate the presence of a single "master gene" for the development of eyes in such distant groups as insects and vertebrates. You can also say this: all the diversity that has a common root testifies to a long journey of trial and error. In the world of invertebrates, the variety of eyes is impressive. They have unicellular, multicellular, straight, inverted, parenchymal, epithelial, simple and complex eyes. One animal can have different types of eyes: wasps have two compound eyes and three simple eyes (ocelli), scorpions have 3-6 pairs of eyes (one pair is main or medial, the rest are lateral). In evolution, compound eyes evolved through the fusion of simple ocelli. All these complex combinations are a clear demonstration of evolution and the search for an adaptive path.

In higher mammals, the variety of eyes is not big. The conclusion suggests itself: evolution at some point moved from experiments with an external perception apparatus to the development of a computational function. Even a small lens and a not too complicated retinal matrix can provide sufficient data flow. The question is the efficiency of processing. An analogy from artificial signal processing systems: a modern smartphone with a tiny lens and a very modest matrix compared to a professional camera can take high-quality picture thanks to its good processing power and software.

An efficient network topology plays an important role in the computing function. In vertebrates, the transition occurs in the chiasm at the level of the central pathways. There are also different variants of such a scheme. For example, in some bird species there is a division of the visual field into halves, when the retina perceives the ipsilateral half located on the same side. And then the intersection is simplified, as in the above diagram.

In primates and humans, the retina of each eye sees both sides of the visual field. In this case, the retina of one eye is divided into the nasal and temporal parts, working with different halves of the field. The nasal ones see the ipsilateral one, which means that their streams must be reversed. This is what happens: the nasal streams intersect in the chiasm. The left half of the visual object is displayed on the nasal (inner) half of the left retina and the temporal (outer) half of the right retina, i.e., these areas see the left visual field. Thus, the left visual field sees the nasal half of the left retina and the temporal half of the right, and the right half of the visual object sees the temporal half of the retina of the left eye and the nasal half of the right. Some of the axons of ganglion cells cross, and some do not, depending on the localization of the cells. For example, one axon from the left retina can go through the left optic nerve, the left side of the chiasm, the left optic tract, and terminate in the left half of the brain. Another axon, also from the left retina, goes through the left optic nerve, passes into the chiasm to the opposite side, then goes as part of the right optic tract and ends in the right half of the brain. Non-crossing axons belong to the ganglion cells of the temporal (outer) half of the retina, and the intersecting axons belong to the ganglion cells of the nasal (inner) half of the retina.

As a result of this organization, the left visual field is represented on the right side of the brain, and the right visual field is on the left. This is an anatomical view of the optic pathway. Now let's look at the information flows:

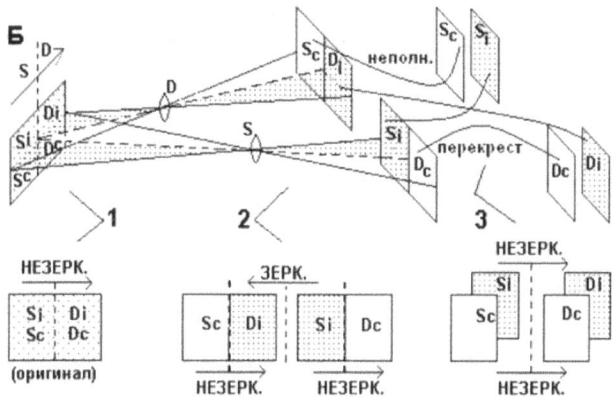

Voronkov, 2009

S and D — left and right eyes. Sc and Si, Dc and Di — the left contralateral and ipsilateral sides of the combined visual field (CVF), the right contralateral and ipsilateral sides of the CVF, as well as the corresponding models. Spot filling — models of the ipsilateral eye of the side of the CVF. Arrows — the direction of the horizontal axis of the coordinates of the CVF. The dashed vertical is the axis of symmetry dividing the CVF and its models in half. As can be seen from the diagram, as a result of a simple combination, the information flows are similar to the original signal in their spatial distribution.

One of the theories explaining such a crossing of neural pathways in the modalities of perception in vertebrates is the somatic twist hypothesis. It suggests that in evolution there was a flipping of the body, when living systems switched from the ventral (anterior, abdominal) location of the spine to the dorsal (posterior). Invertebrates do not have a crossing. But some invertebrates do not have a nervous system at all (for example, sponges), and many exhibit an anatomically diffuse nervous system without a central nervous system (for example, jellyfish). But this does not mean that they do not have intersections of flows and signal distributions. Signal processing can be carried out by physiologically different systems. Functionally, this is the same PAAL scheme.

Anatomical decussation of the optic pathway creates an ordered structure of the flow of information, and at the same time filters-integrators do not have to be topographically consistent with the environment map. Since the brain is an active structure for creating a model of reality, and not a passive receiver-mirror, the location of the population of neurons that make up the integrator in a given functional specialization is not important. It makes sense to facilitate the integrator's work by precursory combination of flows by type of signal or area of the environment into a single stream. But it is not at all necessary for the map of the brain to correspond to the map of the environment. There is no need for the right side of the world to be perceived by the right side of the brain.

As physiologist Nikolai Bernstein wrote wittily: "After all, the central telephone exchange does not have to take care that the switching jacks of subscribers from the northern and southern parts of the city are located accordingly at the northern and southern edges of the switchboard" (Bernstein, 1966).

The brain map reflects the technological chain of the process of creating meanings. It represents the topography of the elements of the system, not the topography of the world. However, the meanings that make up the model of reality as a map of the world must correspond to this reality.

Chapter 3

Signal Identification Technology

A map is not the territory it represents, but, if correct, it has a similar structure to the territory, which accounts for its usefulness.

Alfred Korzybski

The Mind creates a model of the world consisting of representations as data sets about signals. Theoretically, a signal and a set of data about it can coincide. But the brain is a practitioner that does not have unlimited energy, time and space to pursue the impossible and pointless mission of creating a map equal to the territory. It does not copy a signal but reconstructs it. For that, it needs to measure the important parameters and identify the signal.

These are two aspects of the same task of signal reconstruction, but they come from different sides of the PAAL algorithm. Measuring a signal is a perceptive stage of encoding. It determines the quantitative side by answering the question of how much of this or that parameter is in the signal. But the brain cannot know what parameters are important if it does not have an idea of what is being measured. The reverse task of signal reconstruction is about identifying the signal. It answers the question of what kind of a signal is perceived by creating and projecting representations. Of course, it is impossible to answer the second question without dealing with the first one. But without answering the second the answer to the first one does not make sense. As we have noted earlier, making sense of the world is the fundamental function of the Mind.

But it seems that we are facing a paradox: to qualify a signal the brain must have a prior idea about this signal. Almost fifteen hundred years ago, Plato tried to solve this dilemma by suggesting the existence of absolute ideas that are "not born and not perishing, not receiving anything into themselves from anywhere and not entering into anything, invisible and not felt in any way, but given to the care of thought" (Plato, Timaeus, 360 BC). The origin of these ideas was somewhere in the clouds. When they fall from the sky and are given to the care of thought, ideas turn into eidos that are "shadows on the cave wall" of the Mind.

In modern language, it means that projection of the representations that qualify the introjected signals comes not from the brain but from some "cloud storage." The name of modern online storage of information alludes to Plato's theory. But in fact, this technology contradicts his hypothesis, as these storages are quite material computer servers connected to a network, and information (ideas) is the result of the material process of signal transduction within these systems. In Plato's model, ideas are immaterial and just exist forever without being born or modified. How they are given to the care of thought and turn into eidos of the Mind remains a mystery.

This hypothesis seems to solve the initial dilemma but produces further paradoxes that lead to an impasse. To get out of it, we need to look at the process from the height of our current knowledge. Our ideas are not ideal but material, that is why they can develop and change. Paraphrasing Plato, we can say: they can be born and can perish, they receive signals from the material world that enter our Mind when it gives them to the care of our thought. They are saved not in some immaterial cloud storage but in the material substrate of the neural network that constitutes the signal-processing server of our brain.

But how does the brain get the initial prior representations that it projects for identifying signals? Philosophers of the past did not know about the existence of a genetic way of transmitting information. In the absence of knowledge about a process, we tend to think that things just happen in some miraculous way, are heaven sent or literally fall from the sky. But now we are not surprised that the physical features of an organism are inherited and can be changed in the epigenetic process that is encoded in genes to be passed on to the following generations. Should we be surprised that representations of the features of the world can be transferred?

Any living system can project immediately upon entering the world a model of reality that is inherited from all of the previous generations of all life forms. It develops this model throughout life and transmits it to further generations. Of course, the question arises: where does the initial model come from? If we do not resort to magical thinking about ideas being heaven-sent, we should rely on empirical evidence and think in physical terms. The current data shows that even the simplest organic forms exchange substances with the environment. Without any stretch, we can say that they create meanings about incoming streams. As neuroscientist Francisco Varela noted, for bacteria, sucrose solution means "food"; without bacteria, it is just a stream of chemical elements (Varela, 1997).

So, introjection is the initial stage of the signal transduction process in the evolutionary sense. The primacy of projection is algorithmic. To continue the example with sucrose, we can say that for humans it also means food. Our Mind already knows it thanks to the inherited model that starts from the simplest forms of life and projects these qualia onto substances that contain sucrose. We have considered the benefits of the primacy of projection of the PAAL algorithm in the previous part of the study and will return to them more than once.

Now let's get back to the issue of signal identification technology. From this perspective, the brain must find a solution that combines the model and the input data. To achieve this goal, it has to project the model on the data and find intersections. Ideally, this process should go to the point of the complete match. But should time and energy be spent to reach this ideal? The problem of signal identity can be solved approximately and within tolerable discrepancies. Living systems do not need an exact reconstruction, but a level of accuracy sufficient for survival. They do not build a copy of the world, but only a map useful for navigating the world.

To illustrate this, we can again turn to analogies with artificial technology. Let's take an example of a service that allows you to determine the currently playing piece of music: Shazam. In essence, it is a signal identification technology. How does it perform such a function? Here is an ironic picture that represents the naive level of understanding:

For all the naivety, the above picture hits the spot. Of course, there are no live operators guessing the melodies and answering millions of Shazam users' requests. But, as surprising as it may seem, for listening and recognizing the melody our brain uses the same technology as Shazam does. Let's investigate this assumption.

Here is a short functional and technological definition of Shazam: it is a system that transduces signals for the purpose of identifying them. The technological chain is the following: the system perceives the waves of the incoming signal, converts them into patterns of the internal code, compares them with the database and produces a representation. This description is very close to the functional, physical and technological definition of the Mind within TTT. The stages of the Shazam algorithm fully coincide with the PAAL algorithm with the iteration of projection and introjection. The only difference is that a living system creates its own model of reality and Shazam gets its database from the developers. We have to remember the fundamental differences between artificial and natural intelligence not to confuse analogy with equality (see "Part Three. Music of Life").

For a more detailed description of the Shazam technology, let's read the article by the creator, Avery Li-Chung Wang, called "An Industrial-Strength Audio Search Algorithm." He starts by formulating the challenge for the developers: "The algorithm had to perform the recognition quickly over a large database of music, and furthermore have a low number of false positives while having a high recognition rate" (avery@shazamteam.com).

Then he briefly describes the steps of the process:
1. A database of music "fingerprints" is created.
2. The database and the received signal are processed based on the same principles of analysis and transduction.
3. The application, based on signal processing by the device on which it is loaded, sends a fingerprint to the Shazam service, and it looks for matches in the database.
4. If a match is found, information about the song is displayed to the user. Otherwise, an error is detected. The process can be started over.

The "fingerprints" in the database must have a high degree of robustness so that the incoming signal can be identified even with a high level of noise and distortion in the signal itself. They should also have high informational entropy to minimize the likelihood of false positives. On the other hand, too high entropy leads to poor recognition in the presence of noise and distortion. Thus, there should be a balance of high and low informational entropy. It means that the representations must be sufficiently specific to fulfill the function of determining the signal, but not too specific as the risk increases that they will correspond to a large number of signals.

It sounds counterintuitive, but it is true: if the sample is small, then it can match a very large number of signals. A simple example: typing one letter in a search engine will lead to a vast number of matching results. The set of letter combinations will make the output narrower. Typing a word set will give an even smaller choice. You can even type a whole page of text; then you can get exactly to the file you want. But this is not effective. It is easier to restrict ourselves to balanced informational entropy and then choose among a relatively small number of result options.

This is the task facing the living brain: to determine the correspondence of the signal to the stored pattern. The sample should be accurate enough but not too accurate. Otherwise, the chance of type I errors (false positive) increases, and the costs of creating a representation and filtering the comparison results increase. But the accuracy should be so high that it does not produce type II errors (false-negative response when the signal is ignored). Besides, the accuracy and clarity of representation should make it possible to distinguish and identify the incoming signal against the background of other signals. The system must cope with the "cocktail party effect": to highlight the desired signal from the environment of others, even very similar signals. The developers of Shazam faced the same task: the song should be identified among other sounds of a similar frequency range (voices, street noise, and even other music).

What technological decision did the authors make? They settled on an algorithm for extracting peaks in the spectrogram. If there is a specific segment of the spectrogram of the signal along the time axis, which has a high-frequency energy saturation around some center, it will be a candidate for the "title" of such a peak. Amplitude is also important, as there is a logical assumption that the signal with the highest intensity can survive amid noise and distortion. Although for the system to detect even a quiet signal among the background, the frequency patterns and their peak densities are important.

How does our brain cope with the same task? We can purposefully isolate an individual conversation from all other voices, music, rattling dishes, and other background noises during a real cocktail party. Not so much because it is louder than others, but more due to the selection of spectrogram peaks of the signal and comparison with the representation existing in the reality model. We can distinguish the voices and other familiar signals from the general background, which becomes a cacophony for us at this moment since the system directs the computing resource to a specific signal. The rest cease to be coherent patterns, acquire the features of white noise with a smeared spectrogram. But they can become quite coherent at any moment when the brain begins to compare their peaks of the spectrogram with representations present in its model of reality.

It is the so-called figure-background effect, which everyone knows from personal experience and described by psychology at the beginning of the twentieth century (studies of the gestalts). This was only a description of the external manifestation of the system's work. Moreover, they were described as properties of perception. But once we approach the phenomenon from a technological point of view, it becomes clear that perception is only part of the process. You can't select a figure from the background if you don't know what to choose. You need a projection of an existing representation.

Imagine the noisiest "cocktail party" — a colony of seals. Moms and dads, who swam in search of food, return and look for their cubs. They do not see the cub, as it is somewhere far away among the general mass. The main criterion for them is the voice. How do they define it in this cacophony? For us, this is noise because, in our model of reality, the representation "cry of a seal" is very general. It is approximately the same as if Shazam looked not for a specific song, but qualified it as "music/not music." But the brain of a mother seal works like a very accurate Shazam: it can identify the voice of her cub among thousands. How?

The answer seems simple: her brain projects a representation of the voice that identifies the original signal with the necessary and sufficient degree of accuracy. We can formulate it in informational terms: the representation has a balanced ratio of high/low entropy to avoid errors of the first and second types. Because otherwise for the offspring, and, therefore, for the species as a whole, the question of life or death will arise. Rather, it always stands, and it must be solved in favor of life. The coherence and fidelity of the projected reality model are the number one priority in this game. What technological solution provides a chance to win? The answer to this question is not simple anymore.

How does Shazam technology deal with the issue of accuracy of representation? A complex spectrogram of any signal can be compressed to a necessary and sufficient set of key features as a "constellation map":

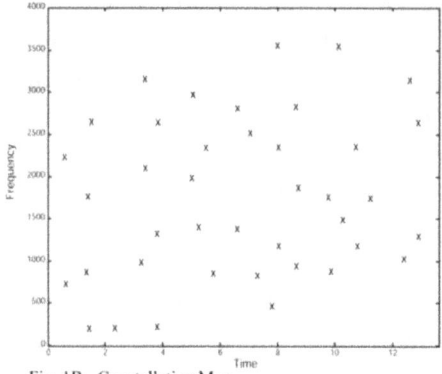
Fig. 1B - Constellation Map

For representing a signal, the entire spectrogram of the development of the signal is not required. The Shazam system does not have to store the whole universe of music, just as the brain does not need to keep the entire universe of environmental signals. Wang notes: "The number of matching points will be significant in the presence of spurious peaks injected due to noise, as peak positions are relatively independent; further, the number of matches can also be significant even if many of the correct points have been deleted. Registration of constellation maps is thus a powerful way of matching in the presence of noise and/or deletion of features. This procedure reduces the search problem to a kind of "astronavigation," in which a small patch of time-frequency constellation points must be quickly located within a large universe of points" (Ibid).

Let's expand the astronavigation analogy to explain this statement. In a real 3D space, the stars may be far away from each other and not related in any way. But for the naked eye, the starry sky appears as a 2D flat canvas where stars are close and even seem to be gathered in some patterns. Today some people think that combining them in constellations was just a whim of ancients who looked at the sky and let their imagination run wild. But all Pisces and Scorpions of the starry sky initially had a practical purpose. Ancient travelers created star maps, where they distinguished anchor points to which neighboring points were associated. This is how the "landmarks" in the sky were created.

Now let's imagine a real sky and real weather conditions. It is not always possible to see anchor points of even very bright stars. But you need to navigate. What should you do if there is a risk of noise and/or deletion of discrete features? Combine them into Aquarius, Virgo, Taurus, and so on. If some points-stars are not visible, then you can, firstly, complete the constellation with a smaller set of points, and secondly, you can correlate some constellations with others even in the absence of anchor points.

The lives of travelers depended on these constellations. They were so important that people, as is often the case, gave them the status of special entities. Gradually, people began to associate these "entities" with their fate as a metaphorical journey along the road of life. They created cults of worship with different names. The most famous is astrology, which originates from the Sumerian-Babylonian astral myths. It may be more ancient since people have been traveling the Earth for a long time, and at night the stars are sometimes the only guiding sign. They really determined fate not as a movement throughout the whole life, but as a specific movement from point A to point B.

Here is how Wang describes the matching procedure of the Shazam algorithm: "If you put the constellation map of a database song on a strip chart, and the constellation map of a short matching audio sample of a few seconds length on a transparent piece of plastic, then slide the latter over the former, at some point a significant number of points will coincide" (Ibid).

This is again just an analogy to illustrate the process. It clearly shows that projection and introjection should overlap each other, and at some point, a "miracle" will happen: the representation in the model will coincide (not completely but in key parameters), and it will give the result that is expected of it in accordance with its function — the signal will be identified and reconstructed.

The constellation map must be compiled with regard to the optimal ratio of low/high information entropy, or otherwise, high/low certainty. The constellation points have low entropy: they are very definite, narrowly focused across the spectrogram range. This can lead to false positives as well as slowing down of the identification process. To solve the problem, the developers used constellation map indexing. Sets of "fingerprint" hashes are created in which pairs of points are associatively combined. Anchor points are selected with a coverage area, where the pairs of points have their own frequency components and differences along the time axis:

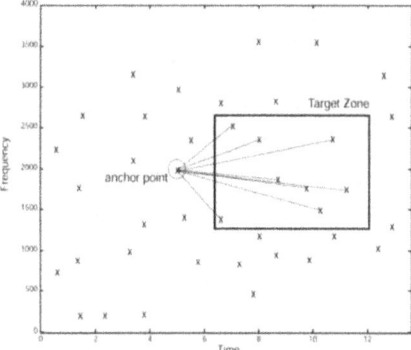

Fig. 1C - Combinatorial Hash Generation

These hashes are reproducible even in the presence of noise, distortion and signal compression during encoding. And each set can be independently coded as a discrete pattern. It can be placed according to the system clock time scale relative to the anchor point. When creating a database, such an operation is performed for each track to form a corresponding list of hashes and their relative position along the timeline. When searching for a song, the fingerprints are used as keys in a hash table. The keys correspond to the time values when a set of frequencies appeared in the music piece, for which there is a model and an identifier-representation (song title, artist name, and so on).

It may happen that a section of song A sounds precisely like a section of song E. Whenever the system finds a matching hashtag, the number of possible matches decreases. But it is very likely that this information will not allow to narrow down the search range to stop at the only correct song. A fragment of a song can be from any part of it, so it is impossible to directly compare the relative time inside the recorded fragment with the database. But if there are multiple matches, the system can analyze their relative timing and thus improve the reliability of the search. This makes it possible not to care about which part of the song the recording falls on.

In addition, the introjected signal includes a lot of noise, which will lead to some discrepancies in the comparison. Therefore, instead of trying to exclude from the matching list everything except the correct composition, the system sorts the records where the overlaps were found at the end of matching with the database procedure in descending order. The more matches, the more likely it is that we have found the desired composition. Accordingly, it will be at the top of the list. The system works probabilistically. There is always an estimate of the probability weights when comparing the database and the input signal. The system cannot afford to wait for an exact match simply because it is impossible by definition: the map is not equal to the territory.

Associative linking into constellations and a probabilistic approach give an enormous acceleration of the signal identification process. As the author notes: "For example, if each frequency component is 10 bits, and the Δt component is also 10 bits, then matching a pair of points yields 30 bits of information, versus only 10 for a single point. Then the specificity of the hash would be about a million times greater, due to the 20 extra bits, and thus the search speed for a single hash token is similarly accelerated" (Ibid).

Of course, not everything is so rosy, since due to the coincidence of hashes, additional processing and screening are needed, which increases the time. And the more extensive the database, the longer this time. For a small base of several tens of thousands of tracks, Shazam can show speed within milliseconds, but for millions, it is seconds. Besides, a large number of points in one zone carries the risk of a combinatorial explosion of pairs (an increase in the fan-out factor), which will require extensive memory resources. Combinatorial associative connections of hashes make it possible to speed up the process by 10,000 times with an increase in the amount of memory only 10 times and leads to a small loss of the probability of determining the signal. The system strikes a balance between specificity and generalization of representation, between low and high entropy, and between processing speed and computational energy costs. Otherwise, it would not have been able to fulfill its function of determining the signal and, therefore, have a commercial success.

We can assume that the normally functioning brain keeps this balance to ensure the correct identification of the signals and the success of the organism in life. The expanding model of reality requires processing and storage capacities, i.e., an increase in the number of elements and structural levels. This is what happened in the evolution of the nervous system. But the larger the volume of representations, the greater the risk of false positives. Phantom representations, as errors of the first type, are the price to pay for the increase in computing power. These errors carry risks and additional costs for correction, but they are justified by the fact that, on the whole, the system has a greater computing resource and better copes with errors of the second type (missing a signal). It means an expansion of the range and volume of the reality model and, consequently, an increase in adaptability.

A living system is forced to balance the desire to expand the range of signal processing and the associated costs and the possible errors with increasing complexity. The system tries not to allow a false-negative response in principle and keeps the risk of a false-positive at a minimum level. To cry wolves when there are none or not to notice that wolves are around are risky options. First means that the system produces a phantom model which is a road to maladaptation. Second means missing a signal, and that could be an immediately fatal mistake. A normally functioning brain corrects errors in the iterative process of comparing projection versus introjection. In the case of systemic violations of the technological chain of the PAAL algorithm, leading to projection and introjection discrepancy, errors in the signal identification and reality modeling process become chronic and threaten the survival of an individual (more on that in "Part Eight. Dissonances of the Mind").

Based on the signal transduction model offered in the previous part of the study we can assume that the brain technology of matching projection and introjection flows is in many ways similar to Shazam. But TTT offers hypotheses that also show differences. They are mostly about the physical implementation of the algorithm.

In the picture with the hypothetical Shazam operators listening to music, one of them exclaims: "I think it's Prodigy!" The brain of an operator trying to guess a melody would perform almost the same signal conversion operations as artificial devices involved in Shazam technological chain. But there are some fundamental differences.

Shazam has a vast database of millions of tracks that a live operator would envy. But it does not have the flexibility as the ability to tolerate signal deviations from representation. Suppose we arrange a competition between Shazam and a fan of Prodigy. In that case, the latter will not care whether it is a studio recording or a concert, which may differ from the first one significantly. Shazam will give a false-negative response or simply will not give any result. The living listener will not experience any difficulties. Moreover, if someone sings or whistles a piece, a Prodigy fan will identify the track, even if the key is different from the original or the rhythm and tempo are not strictly same.

Shazam needs coincidences, not similarities. If you sing "Happy birthday," it will not produce the result. It may have the constellation map of this song, but not performed by you. It has another spectrogram with the other set of combinatorial hashes due to voice timbre variations. Shazam is not guessing the melody. It is looking for a coincidence with a specific dataset. An artificial system is looking for a specific roar of a concrete tiger. The living Mind is satisfied to know that it is a tiger roaring. What do you think is more important for survival?

Wang described the Shazam encoding technology like this: "The pattern of dots should be the same for matching segments of audio ... Each hash can be packed into a 32-bit unsigned integer" (Ibid). The model and the introjected signal are compared in discrete form. For the Shazam system, the input signal becomes a digital code, and the stored representation is a discrete set in a hash table. For a result as the reconstruction of the signal, they must match. Shazam does not look for the perfect match but selects the maximum one. This probabilistic approach is similar to the living Mind. But with this technology, where projection and introjection converge as discrete codes, the variability space is minimal.

In the previous part of the study, we have already discussed the advantages and disadvantages of the digital and analog aspects of signal processing. In short: the digital domain is accurate, while the analog part is efficient and flexible. The brain as a hybrid analog-discrete technological device combines advantages and compensates for the disadvantages of both aspects (see "Part Four. Algorithm of the Mind"). This is the secret of the living Mind in its game of "guessing the melody" of life.

But how does it work physically? We have to get back to the hypothesis proposed in the previous chapter about matching projection and introjection as reference waves superimposed on the object waves. It is more analogous to holography than to the matching of digital samples in Shazam until a significant number of discrete points will coincide. Waves can couple within the freedom of the sync region and in different variants of frequency and phase ratios. The wave nature of representations makes it possible to provide flexibility and resilience. The wave processes are so important for the functioning of the technology of the Mind, that we will be discussing them in detail in the following chapters of this volume and in the next parts of the study.

Here we get back to the question of signal reconstruction technology of the brain. As an example, let's take the research into the workings of a living Shazam — the human auditory modality. The report on the experiment is called "Reconstructing speech from the human auditory cortex" (Pasley et al., 2012).

In the beginning, the authors describe the sound processing performed by the brain: "The early auditory system decomposes speech and other complex sounds into elementary time-frequency representations prior to higher level phonetic and lexical processing. This early auditory analysis, proceeding from the cochlea to the primary auditory cortex (A1), yields a faithful representation of the spectro-temporal properties of the sound waveform, including those acoustic cues relevant for speech perception, such as formants, formant transitions, and syllable rate. However, relatively little is known about what specific features of natural speech are represented in intermediate and higher order human auditory cortex" (Ibid).

Primary perception is well studied. There is little doubt that receptors transduce wave signals into internal code patterns. From the TTT perspective, it is the ADC part of the hybrid chain and the introjection stage of the PAAL algorithm where converting filters are involved. What is less evident is how these encoded patterns turn into signal representations at the higher stages of processing. Classical terminology calls these levels "associative areas" implying that here the streams of encoded features of a signal are combined into a full-scale representation. TTT takes a strictly technological approach and calls them intermediate and higher integrators (see "Part Four. Algorithm of the Mind"). Regardless of the terms, understanding how the "Shazam" of the brain guesses the "melodies" of environmental signals requires a technological approach.

The team of the article authors consisted of a neuroscientist, an engineer, a neurosurgeon, a neurologist, and a psychologist. In formulating the goal of the experiment, they spoke in a strict technical language: "One approach, referred to as stimulus reconstruction, is to measure population neural responses to various stimuli and then evaluate how accurately the original stimulus can be reconstructed from the measured responses. Comparison of the original and reconstructed stimulus representation provides a quantitative description of the specific features that can be encoded by the neural population. Furthermore, different stimulus representations, referred to as encoding models, can be directly compared to test hypotheses about how the neural population represents auditory function" (Ibid).

They took a relatively large number of subjects (15 people) who underwent neurosurgical operations and recorded neural activity in the superior temporal gyrus (STG), part of Wernicke's area responsible for the processing of speech, using electrocorticographic ECoG recordings. This is a method of measuring potentials using electrodes applied directly to the open cortex of the brain. The potentials have an order of magnitude greater amplitude than the EEG, and the equipment provides better resolution. It is used mainly in animal experiments since it is difficult to find appropriate subjects among people.

It should be noted that microelectrode technology is one of the direct methods for studying the activity of neuron populations. It measures action potentials and gradation potentials (intermediate changes in the membrane potential). The spatial resolution of the ECoG is millimeters, and the temporal resolution is milliseconds. The resolution of an individual electrode reaches hundredths of a millimeter, which allows, in principle, to study the activity of one neuron. Also, this technology has a very high sampling rate (more than 10 kHz), which allows for measuring the subtle nuances of fluctuations in the neuron potentials. It means that it is close to measuring the actual wave activity of neurons instead of just counting discrete spikes. This registered wave activity can be converted either into a visual representation on the screen in the form of a spectrogram or into another analog representation (for example, an audio file). Thus, using artificial ADC-DAC technology we can reconstruct how the brain ADC-DAC technology reconstructs the external signal.

If we assume that the brain analyses spectrograms of incoming signals when creating representations, we need to do the same: create a spectrogram from the incoming stimulus signals presented to the subjects (in this experiment, it is the sound of certain words), produce a spectrogram from the received signals of neuronal activity and compare result. If there are similarities, then the hypothesis can be considered working and developed further, improving technologies and approaches to the design of the experiment. But apart from all the technical difficulties, the question arises: which paradigm of creating a spectrogram-representation to choose — linear or nonlinear?

A spectrogram is a representation in terms of frequency and time. These are not static parameters. Natural sounds have modulations which are of fundamental importance for signal differentiation (in this case, speech intelligibility). Temporal modulations can occur at different rates and in different ranges of the frequency spectrum. For example, speech is characterized by the slow modulation of the tempo of syllables and fast modulation at the beginning and end of a syllable. It also has a broad spectrum for vowel formants and a narrow range for their harmonics. Different formants (vowels, consonants) have different spectral widths. Within vowels and consonants, there are differences in modulation and spectrum.

So, the spectrogram of the original signal is a complex and dynamic structure. And the question arises: is there a linear dependence of the neural response on this spectrogram? The key requirement of the linear model is that the neural response must coincide with fluctuations in the stimulus spectrogram. The authors tried experiment design according to this paradigm. Schematically it looks like this: the neuron signal passes through the reconstruction model, and a spectrogram is obtained that reflects the signal's main spectral-temporal features, for example, the concentration of energy on the harmonics of vowels and high-frequency components of fricative (sibilant, hissing) consonants.

A linear model is satisfactory in terms of general energy analysis. For that, neural responses have to correspond to certain requirements. First, the recorded zones must be selectively tuned. The authors applied techniques for revealing the selectivity of the response to a specific stimulus and revealed sensitivity to different frequencies. This allowed the obtained spectrogram to reflect the main general energy features of the spectrogram of the original stimulus:

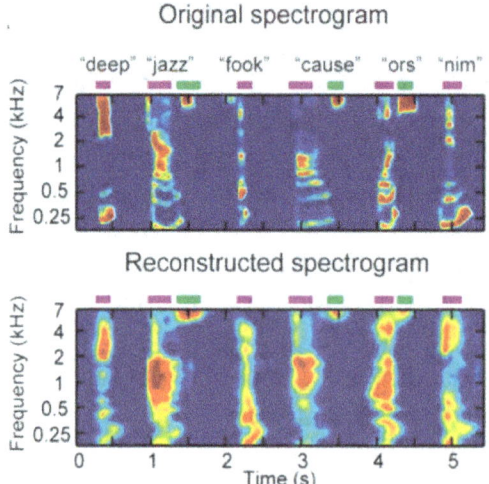

The next requirement for a linear model is a reliable relationship between the response of neurons and fluctuations in spectrogram envelopes. And here, the limitation of the linear model is revealed, especially at a high tempo. Considering this, the authors applied another model: filters of wavelet analysis and transformations of the original spectrogram. It is a method of converting a signal into a form that either makes certain quantities of the original signal more processable or compresses the original data set. It modifies the signal, making it more acceptable to the processing system, but retains its characteristics.

The authors applied filters selective to modulation of the input signal, which analyzed the spectrogram and detected energy in it at a different time and spectral scales (nonlinear operation). They noted: "Reconstructing the modulation representation proceeds similarly to the spectrogram, except that individual reconstructed stimulus components now correspond to modulation energy at different rates and scales instead of spectral energy at different acoustic frequencies ... The nonlinear model yielded significantly higher accuracy compared to the linear model" (Ibid).

The authors showed this graphically as the effect of tempo on reconstruction accuracy:

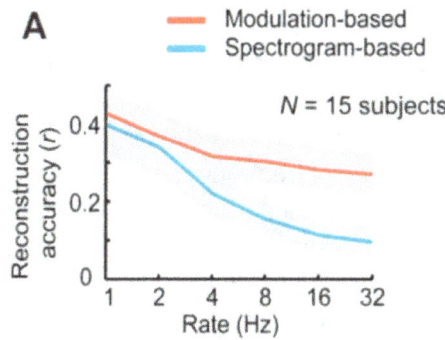

In the bottom graph reflecting a linear model, the accuracy drops steeply. In the modulation-based model (top chart), it is considerably better.

The incoming spectrogram was two-dimensional S (f, t) along the space and time axes, and the outgoing spectrogram was a four-dimensional representation of the energy modulation M (s, r, f, t) along the axes of spectral modulation s, temporal modulation r, frequency f, and time t. Thus, the representation was scaled in time and space, and it became a nonlinear model of the original signal:

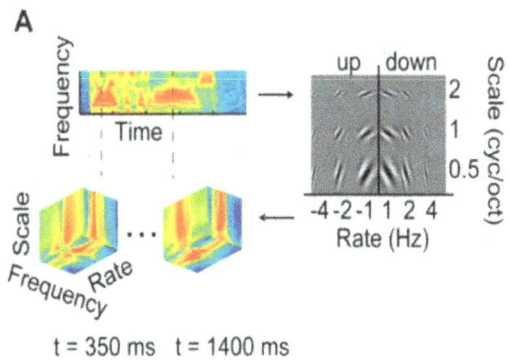

What is the problem with the linear model? "The linear model assumes that neural responses (high gamma power, black curves, left) are envelope-locked and directly track this rapid change. However, robust tracking of such rapid envelope changes was not generally observed, in violation of linear model assumptions" (Ibid). In plain words, the linear model assumes that the Mind is a mirror. But it doesn't work that way. The brain creates representations, not reflections of signals. But this does not mean that the mirror is crooked. There is simply no mirror but a transformation in the filter chain. As a result, the map is not equal to the territory. Waves of the Mind are small wavelets because they are the result of modulation and transformation.

What did the authors of the experiment do to make the resulting representation more accurate and dynamic? Like the brain itself, they made a chain based on the principle of band filtering, piecewise wavelet analysis, and Fourier transform (see "Part Four. Algorithm of the Mind"). Their model made it possible to obtain information in the neural response based on energy modulation. Moreover, such a model showed that with a sharp drop in the original spectrogram envelope's speed, there was no drop in the neurons' activity. There is no linear relationship between the signal and the neurons firing rate. The neural code is not in the tempo of discrete spikes, but in frequencies and phases as melodies and rhythms of the brain (more on that in "Part Six. Harmonies of the Mind").

The authors needed to reconstruct the original signal as a waveform, i.e., to get all the wave characteristics (frequency, amplitude, phase). To obtain the missing data, they used an iterative projection algorithm based on prior speech data. They had an initial speech spectrogram, which remained to be compared with that obtained from the activity of neurons.

The availability of information about the parameters of the wave signal made it possible to reverse the transformation of a complex modulated representation back into a spectrogram. Next, the spectrogram was transformed into an audio file using a speech recognition algorithm to arrange all sounds according to the temporal structure of speech. As a result, the music of the neurons started playing (http://www.berkeley.edu/news2/2012/01/recon_UCB_012112.MOV). Those who expect to hear an exact copy of the words will be disappointed. Articles about this experiment were titled "Neuroscientists have created a computer program that can decode your thoughts" (https://www.extremetech.com/extreme/116447-neuroscientists-create-computer-program-that-can-decode-your-thoughts).

However, this is not only and not so much a computer program, and it is not at all about decoding thoughts. It was about reconstructing the representations of sound signals at the intermediate stage of the auditory modality. But for those who are aware of the complexity of the process, this is a miracle of technology. The authors wrote in a strict technological tone: "In the case that STG applies a highly nonlinear stimulus transformation, an exact reconstruction of the acoustic signal from STG responses would not be possible. Instead, speech reconstruction provides an important tool to investigate the critical features that are faithfully represented at different stages of the auditory system … Although more work is needed to characterize the neural representation in the STG, this suggests that such key features are preserved at this stage in auditory processing. Our results are therefore consistent with the idea of STG as an intermediate stage in a hierarchy of auditory object processing ... But development of higher level encoding models could be required to describe more anterior areas in the ventral auditory pathway. As understanding of cortical speech representation improves, future research into speech reconstruction may also be useful for development of neural interfaces for communication, for example by revealing the content of inner speech imagery" (Ibid).

In such roundabout ways, the authors say that the goal is, of course, mind-reading ("revealing the content of inner speech imagery") and connecting the brain and external devices for transmitting thoughts ("neural interfaces for communication"). But for this, it is necessary to move further along the chain of formation of representations up to the highest integrators. Indeed, at the stage of higher integration, not representations of individual words are created, but speech as a whole with its lexical, syntactic and semantic connections. It goes beyond the auditory functions of analyzing and transforming sounds. But the study of "inner speech imagery" (mind-reading) with the correct conceptual approach sounds not like fantasy, but as a feasible, although technologically and technically not an easy task.

What did the authors do in this experiment from the point of view of the TTT and PAAL model? They investigated the intermediate filters-integrators, where the signal is detached from the original because it was transformed in sampling, quantization, modulation and interpolation process (see "Part Four. Algorithm of the Mind"). It has moved from the world of signals into the world of representations, from the outer to the inner universe. Therefore, the researchers had to repeat what the system itself does: to modulate and then transform the signal into a wave representation and apply the projection algorithm of the existing model of this signal.

It may seem that all these complicated technological tricks are just wishful thinking. They took some cortex area activity, adjusted it to the sounds, and said that they match. And there is truth in this: the experimenter always proceeds from the hypothesis (expectations) and compares the results with it. But the truth is that the chain of the Mind itself is also "wishful thinking," as the projection of a model and comparison with introjection. And in the course of this process, the living system does many transformations and other "tricks" so that the result is close to the expected.

When experimenters go along the technological chain of the signal processing, they, willy-nilly, repeat the PAAL scheme as a technological chain of the Mind. In doing so, they solve precisely the same reverse problem of stimulus reconstruction: "measure population neural responses to various stimuli and then evaluate how accurately the original stimulus can be reconstructed from the measured responses" (Ibid). After all, a living system also measures its own responses to stimuli and estimates how accurately the original signal is reconstructed as a result of all transformations. Let's once again draw attention to the fact that the experimenters, when analyzing signals at the intermediate level of integration, had to use an iterative algorithm for comparing the a priori data set with the input set. They did the same thing that the brain does when processing incoming information: it superimposes the flow of projection of existing representations onto the flow of incoming patterns.

To put it simply, we hear what we expect to hear. The signal must be placed in the existing "Procrustean bed" of the model. The system cannot perceive the signal without any preliminary expectation, without the projection of the reality model in the context of this type of signal. This means that higher integrators project the reality model onto the incoming streams from the primary and intermediate integrators. This is evident even by the timing of activity between these levels of the technological chain of the PAAL algorithm. For example, research on mice shows that stimulation of the orbito-frontal cortex (OFC) in vivo using optogenetic methods induces activity in the primary auditory cortex (A1). Correlating this stimulation with sounds shows that the responses of A1 neurons to sounds change during projective activity in OFC. The authors of the experiment note that "results identify a direct connection from OFC to A1 that can excite A1 neurons at the earliest stage of cortical processing, and thereby sculpt A1 receptive fields" (Winkowski et al., 2017).

Let's take a study of another natural Shazam: the auditory cortex activity in the brain of songbirds. The authors start with a very technical description of the problem: "Birdsong is comprised of rich spectral and temporal organization, which might be used for vocal perception. To quantify how this structure could be used, we have reconstructed birdsong spectrograms by combining the spike trains of zebra finch auditory midbrain neurons with information about the correlations present in song." (Ramirez et al., 2011).

The scientists let the finch listen to the songs of other birds, recorded the neuronal activity of its auditory modality using electrodes inserted into the brain, and combined the processed activity with data about such songs. It is a demonstration of the PAAL in action: the projection of the a priori model, comparison with introjection, and the formation of the current model of the song. In this case, the researchers did part of the work for the finch's brain in the projection part of PAAL: they created a model from a statistical sample of finch songs and projected it onto the input signals processed by the auditory modality.

Here is what the authors reported: "When we evaluated the reconstructed spectrograms in the Fourier domain, we found that these responses do a fair job of reproducing temporal and spectral frequencies … When combined with the joint spectrotemporal correlations of zebra finch song, we found an improvement in the coherence in these regions … These results are qualitatively similar to previous findings showing that the auditory system of zebra finch, as well as other songbirds, can recognize songs even when some of the fine details of the song signal have been degraded by various types of background noise" (Ibid).

Of course, these experiments are not brain code reading. The authors do not say that they have deciphered it. They describe the first attempts at reconstructing brain-generated representations. We can use the following analogy: take a recording of one musician, compose its spectrogram, take a recording of another musician and do the same, then superimpose a database of many performances on it and check the correspondence of the spectrograms of the first and second. To do this, one does not have to know either musical notation or the intricacies of the performance of the piece with its melodic, harmonic and rhythmic structure. At this stage, knowledge of the code is not required. This is a kind of workaround. And it works well because the technological chain corresponds to what the brain does. But to read and reproduce the music of the brain, we have to understand its musical notation (more on that in "Part Six. Harmonies of the Mind").

Let's take the example of visual modality research. The article is called "Bayesian reconstruction of natural images from human brain activity" (Naselaris et al., 2009). What is the goal of the authors? "We demonstrate a new Bayesian decoder that uses fMRI signals from early and anterior visual areas to reconstruct complex natural images. Our decoder combines three elements: a structural encoding model that characterizes responses in early visual areas, a semantic encoding model that characterizes responses in anterior visual areas, and prior information about the structure and semantic content of natural images. By combining all these elements, the decoder produces reconstructions that accurately reflect both the spatial structure and semantic category of the objects contained in the observed natural image ... The goal of reconstruction is to produce a literal picture of the image that was presented" (Ibid).

What was the result? "Our results show that prior information has a substantial effect on the quality of natural image reconstructions. We also demonstrate that much of the variance in the responses of anterior visual areas to complex natural images is explained by the semantic category of the image alone" (Ibid).

The authors need two databases (structural and semantic) because they measured the activity of primary integrators and subsequent ones. The first encode the structure of the signal, the second create meanings as reconstructed signals. In the structural model, the probability is determined by spatial frequency, in the semantic — by the meaning of the scenes in the photographs. The model distributed the neural responses into semantic categories, which were named by the subjects themselves when viewing the pictures before the start of the measurement.

The authors emphasize that the images were not pre-selected to fit into specific categories. On the contrary, categories were created as a result of subjects describing randomly selected photographs. After the photos were categorized, the expectation maximization optimization algorithm (EM) was used to associate a semantic model with each response. Thus, the semantic model reflected neurons' preferences for each category, i.e., tuning of a given neuronal population to a specific pattern. The structural model was the best predictor of the activity of the primary posterior zones, while the semantic model predicted the activity of the zones further ahead in the occipital visual cortex.

The subjects looked at 1,750 monochromatic photographs while being in the fMRI machine. During the reconstruction of the image, the subjects were given 120 new photos. The first information was included in the models, and the second was used to generate direct representations as a result of correlation with models.

First, the authors started with a flat prior model that distributed the same probability weight over all images. This is a null model where only neuronal activity is used for reconstruction. It gave no result. Then they took a structural model. And again, no success. They added a semantic one. The result was better, but still not an acceptable reconstruction. The authors found that a small model of 1,750 images was not enough. This volume provided a model of the activity of brain zones but did not offer a model of the signal itself. They had to use what the authors called the hybrid model: fusion of the structural, semantic model and the a priori model of natural images.

To make the hybrid model work, the authors compiled a database of six million natural images selected at random from the internet. In other words, they have done what all living systems do: they have accumulated enough data for the necessary accuracy of signal identification. And the larger this dataset, the better the result. The accumulation of statistical information about signals in the reality model takes place as physical adjustments of system elements to specific patterns. And the more often this pattern occurs in the course of signal processing, the more stable the settings of the impulse responses of the filters, the more stable and precise the representation.

As a result of all the work of creating models and comparing them to signals, the authors got the following results:

The left column is the target image (original photo), the middle is the structural model, and the right is the hybrid one. There are apparent attribution errors in the middle column: the model mistook a group of buildings for a dog, and so on. The third column contains attribution errors, but they are no longer so extreme. Buildings become buildings; people become people; grapes become berries. The model takes the snake for a caterpillar, but even in real life, the brain can make such identification errors. But it is better not to confuse a snake with a caterpillar to live longer. The way out is accumulating information, improving the posterior probability distribution by enhancing the prior one, and increasing the accuracy of the representation.

A hybrid model has worked, as it reflected the convergence of the projection of a priori information and the introjection of current data. In the experiment, the PAAL algorithm again clearly manifests itself. And this happens naturally since a living system (in this case, researchers) cannot cognize, except according to the scheme common to all living systems.

What conclusion do the authors draw from the experiment? They ask: "But is this really reconstruction?" They answer: "Reconstruction using the natural image prior is accomplished by sampling from a large database of natural images. (It) will always correspond to an image that is already in the database. If the target image is not contained within the natural image prior then an exact reconstruction of the target image cannot be achieved. The database used in our study contains only six million images, and with a set this small it is extremely unlikely that any target image (chosen from an independent image set) can be reconstructed exactly. However, as the size of the database (i.e., the natural image prior) grows, it becomes more likely that any target image will be structurally and/or semantically indistinguishable from one of the images in the database. For example, if the database contained many images of one person's personal environment, it would be possible to reconstruct a specific picture of her mother using a similar picture of her mother ... Much of the excitement surrounding the recent work on visual reconstruction is motivated by the ultimate goal of directly picturing subjective mental phenomena such as visual imagery or dreams. Although the prospect of reconstructing dreams still remains distant, the capability of reconstructing natural images is an essential step toward this ultimate goal. Future advances in brain signal measurement, the development of more sophisticated encoding models, and a better understanding of the structure of natural images will eventually make this goal a reality. Such brain-reading technologies would have many important practical uses for brain-augmented communication, direct brain control of machines and computers, and for monitoring and diagnosis of disease states. However, such technology also has the potential for abuse. Therefore, we believe that researchers in this field should begin to develop ethical guidelines for the application of brain-reading technology" (Ibid).

The authors say that in order for the correlation to be accurate, a huge database must be compiled, and preferably pertaining to a given individual. It is necessary to compile a subject's reality model to compare the activity of neurons with the received signals from the environment. Even millions of images are surely not enough. This seems to be a somewhat costly and ineffective way, or rather, unrealizable in principle. Moreover, it looks like a cycle of logic: to make a reconstruction of a representation, one must have representations. Or more simply: to read thoughts and images, you have to know these thoughts and images. But maybe it is better to decipher the brain code to reconstruct representations based on its patterns?

Without learning to read, you cannot understand the meaning of the text. You can analyze the texts for a very long time, making correlations between them, but still not understand a word. It is impossible to read the brain using a tool (in this case, fMRI) that says nothing about the patterns of the code. It's like a child that reads a book by looking at pictures. We cannot read the brain without deciphering the neural code. The authors' call for ethical guidelines is a little bit premature as their technology is not about brain reading at all.

But what is important from the technological point of view, it that the experiment confirms the importance of the model projection for solving the problem of qualifying a signal. It works for the artificial signal processing technology used in the experiments and for the natural technology of the Mind. Of course, the authors wanted an exact copy of the stimulus to appear on the screen. It would measure the success of the experiment. But for the brain, creating a copy of a signal is not a measure of success. The measure of success is survival. And this is where the secret of living systems lies: they need not a copy of a signal, but a reconstruction with an accuracy level that is necessary and sufficient for survival. They create a map that is useful for navigating the world but not equal to this world.

CHAPTER 4

TECHNOLOGY OF OVERCOMING PHYSICAL LIMITATIONS

The properties of this external world, universals, must somehow be embedded into the functional workings or neuronal circuitry of the brain. Such internalization, the embedding of universals into an internal functional space, is one of the essentials of brain function.

Rudolfo Llinas

The brain faces an obvious technological problem: the inclusion of a potentially infinite world of signals into a limited volume of channels for their reception and processing, as well as storage in a limited volume of a substrate as a carrier of encoded information. In general, the nervous system's entire evolution can be called a process of overcoming the above limitations. As we have repeatedly emphasized, the emergence of artificial technologies is a stage in the development of the Mind, which fits into this general outline of the desire to overcome limitations and solve a technological problem. The first information revolution (the invention of writing) was also aimed at overcoming them. The modern "information era" is a continuation of the history of the development of the living Mind. But in this chapter, we will talk about the internal technologies of our body.

As a specialized structure, the brain appeared relatively recently and at the moment is the pinnacle of the development of technological solutions in this direction. To understand the essence of these solutions, we have to proceed from the awareness of the limitations themselves. By understanding the challenges facing the system, we can hypothesize how it solves them. Here again, the analogy with artificial systems facing similar tasks will help us. Devices created by our brain for solving signal processing problems help us understand the work of the brain. If the functions coincide, then we can assume that the solutions are largely similar. They may differ in the details of the embodiment since there is a difference in material substrates.

In some ways, solutions in artificial technologies are ahead of the capabilities of living systems, and in others, they are lagging behind. We have already considered some aspects of such a situation in the previous part of the study, and we will return to them in this and subsequent ones. This chapter is not intended to be an exhaustive list of limitations and solutions to overcome them but only highlights some aspects.

So, the environment's signals are potentially unlimited and enter our brain, which has physical limitations. Therefore, they must, firstly, be transduced and modulated under these constraints, and secondly, integrated into the reality model. For this, the system must have channels for perception, transmission, processing, storage and reproduction.

Network channels have two limitations: width and power. They operate in a specific frequency range and are limited in terms of the power of the conducted signal. As a result, both restrictions affect the information capacity: the amount of information that the channel can reliably process in a specific period (number of bits per second). When developing technologies, engineers have channel specifications and design a system that processes the maximum number of bits with specific channel capabilities. But the characteristics of the channels can improve in the process of technology development. The evolution of the nervous system's engineering followed the same two directions: the development of the characteristics of channels and filters; optimization of the system architecture and algorithms of its work.

The representation of each signal with many parameters is a convolution of individual filters operation with a smaller set of parameters. The system works on the principle of "divide, rule and combine." The brain first disassembles the flow of signals into puzzle pieces and then puts them together into a picture of reality. Primary converters are tuned to certain signal parameters and create a stream of differentiated code where each element has a unique path (branch) to form part of the representation. The brain uses local analysis at the receptor and primary processing level, then moves to global analysis at the modulator and primary integrator levels. In intermediate and higher integrators, all flows are synthesized and synchronized into a single model of reality.

Let's take the example of the visual pathway again. We have already discussed the specialization of vision receptors by frequency. But the vital part of the information is contained in the transitions (boundaries, contours). It is necessary to encode these transitions as phases to get a representation with the outlines of objects. For this, at the level of primary transducers in the retina, there is an organization of receptors according to the center-periphery principle. It allows transmitting information about the differences in the parameters of excitation of cells located in the center and at the periphery of the receptive field. Thus, ganglion cells can send information about image contrast to neurons of higher levels. An essential part of the retinal topology is the size of the receptive field, which affects the spatial frequency of visual information: small receptive fields are activated by signals with high spatial frequencies and fine image detail; large receptive fields — by signals with low spatial frequencies and poor detail.

Many signals can have the same power and frequency but come from different sources. Only phase analysis allows determining the position of signals in space and their development in time. However, if the amplitude-frequency analysis is removed, further differentiation is difficult. The outline of the object will be clear, but "all cats are gray in the dark." Moreover, errors (illusions) of perception may occur when one object is mistaken for another if their outlines (phase spectrum) are similar. But in any case, the analysis of the phase, rhythmic structure of the music of the world is the basis for signal reconstruction. If the phases are randomly distributed, the representation will be significantly distorted compared to the original.

In artificial image processing, there is a low-pass filter algorithm called Gaussian Blur moving average (GB) or Gaussian smoothing. It is usually used to reduce noise components and unnecessary detail and highlight the structure of an image on a different scale. Creating a representation by reducing detail may sound counterintuitive, but remember that too much is never good: details prevent you from "seeing the forest for the trees," creating a representation of an object. Since the Fourier transform of a Gaussian distribution will also be Gaussian, this means a decrease in the high-frequency component. The GB impulse response has circular symmetry and is separate in both horizontal and vertical filtering. In a 2D matrix, 1D machining directions can be combined and stacked together.

The topography of the receptors on the plane of the retina has a circular center-periphery organization. Main information is processed in the center (Gaussian distribution), but each filter is circularly symmetric. The fields overlap each other, and even coverage of the visual space is obtained. This allows mixing information from neighboring areas, while maintaining their independence. A balance is achieved between specialization and the unification of the computational efforts of the primary signal processing. The resulting signal representations have minimal border effect.

At the entrance to the system, "coarse" filtering occurs — for example, basic trichromatic (RGB) signal processing. By the way, before the measuring technology made it possible to isolate the spectral sensitivity of three types of cones in the retina in the 1960s, there were many theories and mathematical models of signal processing at the eye level, which attributed very diverse functionality to the first level neurons. Now there is a consensus that they encode the frequency range of light waves. But there is still no agreement regarding the processing of luminance and contrast. Sometimes paradigms diverge to "warring camps." For some reason, the theories of edge position coding and spatial frequency coding are considered mutually exclusive. Multi-channel frequency analysis theorists call the edge theorists "feature creatures," while the latter call the former "frequency freaks."

The history of this struggle of ideas has been going on for more than half a century. In 1959, David Hubel and Thorsten Wiesel conducted an experiment with the Primary Visual Cortex of an anesthetized cat (Hubel, Wiesel, 1959). They inserted a microelectrode into it, projected light and dark patterns onto a screen in front of the cat, and measured the intensity of neuronal activity. Some neurons worked quickly when the lines were at one angle, while others responded best at another angle. Some of these neurons reacted differently to light and dark patterns. Hubel and Wiesel called these neurons "simple cells." Other neurons, which they called complex cells, responded to edges no matter where they were located in the neuron's receptive field and could preferentially respond to movement in specific directions. This work received high acknowledgment and was subsequently awarded the Nobel Prize in 1981.

But in 1968, studies showed that the thresholds for the perception of contrast are determined by the fundamental frequency component of the wave, and the individual thresholds for elements of a complex wave are determined by the harmonic components of its composite sinusoids. Neurons were found to be sensitive to sine-wave gratings — a sequence of changes in light and shade, which has its frequency (stripe width), amplitude (the intensity of the difference between light and shadow) and angular phase. The authors suggested that in the nervous system, there are mechanisms selectively sensitive to specific ranges of frequencies, and "the frequency selectivity of these mechanisms must be determined by integrative processes in the nervous system" (Campbell, Robson, 1968). But this idea remained at the side-lines of the mainstream.

Twenty years later, other researchers put forward a new explanation for the work of neurons in the Primary Visual Cortex — the Spatial Frequency Theory (De Valois, De Valois, 1988). Hubel and Wiesel, in their experiments, of course, correctly recorded the sensitivity of the neurons of this zone to the characteristics of visual stimuli, but their hypothesis-explanation that neurons encode the geometry of lines was too simplified. Neurons encode the waveform characteristics of a signal.

The hypothesis of the authors of the Spatial-frequency theory (SFT) was confirmed experimentally. When presented with a stimulus like a checkerboard or a Scottish plaid, neurons responded to spatio-temporal patterns with certain frequency characteristics and remained calm in response to the individual orientation of the lines.

What about old experiments? The result of an experiment can be interpreted in many ways, based on the hypothesis (the projection of the model is primary). Hubel and Wiesel showed various simple lines to the cat, and its neurons responded to them differently. It is a fact. But who said that neurons encoded lines as such, literally like pieces of reality, and not the amplitude-frequency and phase characteristics of these lines as wave signals? Any signal, whether it is a simple line or a complex object, can be processed and its representation created by encoding the characteristics of the light waves by decomposition into composite sinusoids.

SFT was a hypothesis of a higher order, encompassing phenomena that the previous one could not explain but not contradicting it. For example, in an experiment on monkeys, it was shown that the receptive fields that perceive edges and lines are tuned to a narrow frequency range (De Valois, Albrecht, Thorell,

1982). The authors wrote: "If a complex stimulus is to be analyzed into elements along some dimension, one clear requirement is units which are highly selective (that is, narrowly tuned) along that dimension. In an analogous situation in the auditory system, for instance, units in the cochlea are very selective to temporal sine waves of different frequencies" (Ibid).

De Valois and colleagues' experiments showed that cells respond not only to the orientation of stripes and edges, as has been assumed since the experiments of Hubel and Wiesel but to the orientation of the spectrum of each pattern and its fundamental frequency. Neurons work as detectors of signal wave characteristics, as analyzers of spectral information. The neurons of the Visual Cortex, as primary integrators, perform Fourier-type transform, specializing not in geometry as such, but in frequency, amplitude, and phase characteristics reflecting this geometry. Composite sinusoids, which are obtained during the analysis and decomposition of the original complex signal, are again combined in the process of a Fourier-type transformation, integrated into a complete and coherent picture in subsequent integrating filters. Thus, we can see the world not as a collection of discrete lines and geometric shapes but as a continual picture. Decay into simple geometry occurs in the event of a malfunction of the higher integrators (more on that in "Part Eight. Dissonances of the Mind").

The authors of SFT limited themselves to attempts to explain the process of creating representations in visual modality at the level of primary zones. But even at this level, they faced misunderstandings. There was only one problem with the authors' explanation: not everyone understood what they were talking about. Few physiologists and psychologists spoke the language of physics and mathematics.

When Hubel and Wiesel said that neurons encode the lines, edges, and other attributes that make up the picture of the world, everyone understood. It's simple: a neuron encodes "pieces" of reality, and then all these "pieces" are combined into a single picture. The explanation is clear, although it does not answer the central questions: how are the pieces encoded, and how are they integrated? They just imagined that everything is combined somehow and celebrated the discovery.

Here is what the authors of the SFT wrote: "It sometimes appears that the resistance to accepting the evidence that cortical cells are responding to the two dimensional Fourier components of stimuli [is due] to a general unease about positing that a complex mathematical operation similar to Fourier analysis might take place in a biological structure like cortical cells. It is almost as if this evoked for some, a specter of a little man sitting in a corner of the cell huddled over a calculator. Nothing of the sort is of course implied: the cells carry out their processing by summation and inhibition and other physiological interactions within their receptive fields. There is no more contradiction between a functional description of some electronic component being a multiplier and its being made up of transistors and wired in a certain fashion. The one level describes the process, the other states the mechanism" (De Valois, De Valois, 1988).

Indeed, neurons are not mathematicians, just like the resistors in a calculator circuit are not mathematicians, but they perform a function that can be described mathematically. Mathematics is a language with its own set of symbols, a way of describing the world. It does not say how something should be but explains how something could be. If the mathematical apparatus called Fourier transform describes something with a greater degree of accuracy and with a wider coverage of the observed phenomenon, it is at least unreasonable to abandon it.

The mathematical formalism of Fourier analysis describes how any pattern in space-time (signal dynamics) can be transformed into a set of wave structures that encode the frequency, phase and amplitude of the signal. The reverse operation creates a model of the original signal, restores it. Different representations of various signals in the form of such wave patterns can be integrated into a single coherent structure based on the mechanisms of frequency-phase synchronization. Such processes are a physical reality, and mathematical equations are the way to describe them.

In mathematics, the properties of the relation between a signal and its representation are described by the Duality theory (Pontryagin duality). Here are its postulates in short form: complex periodic functions can be transformed into a Fourier series and can be recovered from this transform; complex functions in their real part have a Fourier image and can be reconstructed from it; complex functions of a finite set have a discrete Fourier transform and, again, it can be reconstructed from the image.

Lev Pontryagin had nothing to do with the study of the brain. He was interested in mathematical analysis, oscillation theory and control theory. But the brain is engaged in analysis and control, and its physics is oscillatory. If we look closely at the postulates of the duality theory, it becomes evident that they represent a capacious description of the process of creating a reality model by the brain. Complex functions of continuous signals of the environment can be transformed into frequency components (Fourier analysis) and finite sets (discrete Fourier transform), which leads to the creation of an image (representation), according to which the original signal can be reconstructed in the necessary and sufficient degree of accuracy.

One of the main arguments against "frequency freaks" is that the description apparatus they use is counterintuitive (read — incomprehensible). However, everything is precisely the opposite. It is so intuitive that it's incredible how it hasn't been used before to describe processes in the brain. It is more complicated than the linear formulas for summing "shots" of neurons, but it is quite accessible and practically applicable to describe nonlinear wave processes.

There are many very intuitive analogies of phenomena that can be described using the Fourier apparatus. But for me, the musical analogy is the best. Any complex symphony consists of various parts, which in turn consist of simpler parts up to the most basic frequency-rhythmic elements (notes). As a complex function, the symphony is composed of them; it can be decomposed into them and reconstructed according to them. It has its image in them and can be reconstructed from this image.

Is this description counterintuitive and incomprehensible? But such a description word for word repeats the basic postulates of Pontryagin duality theory, which was a purely mathematical apparatus of Fourier transforms in the theory of topological groups. It looks interesting, right? The same instrument is applicable in seemingly completely different fields of knowledge. But it is not a random coincidence. Various areas of knowledge can talk about the same physics of the process. The same physical phenomena can be described in different ways, but they can be described in the same way. But in the first case, an illusion is created that they are different, and in the second, the "veil" falls, the fog dissipates, and something becomes more transparent and understandable.

Music staff is the spectrogram with space-time measurement axes. Notes are the simple components of a complex spectrogram with frequency (pitch), amplitude (intensity) and phase (sequence and duration) characteristics. A complex symphony consists of them physically, and it can be described graphically as a coded representation using a specific symbolic apparatus (notation). It can be represented graphically as a spectrogram of all these components or mathematically utilizing Fourier-type transforms. It will be an integral (a complex sum of small parts), where the main variables will be space and time, frequency and rhythm.

$$\hat{f}(\omega) = \frac{1}{\sqrt{2\pi}} \int_{-\infty}^{\infty} f(x)e^{-ix\omega}\, dx$$

This basic formula for Fourier transforms is a way of describing the integration of frequency and temporal parameters in mathematical symbols. As a staff and notation on it is also a form of expressing the integration of frequency and temporal components of a complex symphony by signs of the musical code.

Let's return to the coding of signals by the visual system. As we have already noted, the hypothesis that neurons do frequency and phase analysis of the signal does not contradict what Hubel and Wiesel found. Moreover, it explains the phenomenon: neurons reacted to the edges and stripes of light and shade precisely because they encode the spatial and temporal frequency of the light signal.

At the input level, filters work as LPF, i.e., they filter out large variations of the signal (this applies to both chromaticity and luminosity). Receptors of the eye are tuned to a specific frequency range and the perception of transitions due to the spatial organization of the center-periphery with on-center and off-center cells (on-bipolar and off-bipolar cells). A neuron tuned to a particular spatial frequency will respond to a contrast edge passing through the center of the perception field. And vice versa, if it is tuned to a specific contrast bandwidth, it will also have a certain band-pass tuning curve.

At the level of integrators, the signal structure is processed more subtly, and more detailed representations are created. This system design optimally solves the cost/benefit balance. In the perception organs, the number of neurons is small in relation to the signals of the environment and to the number of neurons at the subsequent stages of processing. There should be a technology that allows to "pack" the maximum amount of information into a relatively small number of neurons. One of the indications that coarse filtering occurs at the level of primary signal processing is the fact that the visual perception of an infant differs from that of an adult in terms of detail. Experiments have shown a significant contrast sensitivity function difference between infants at three months and adults, suggesting differences in spatial frequency and phase processing. At this age, the infant has a sufficiently developed binocular vision and primary processing apparatus, but the integrators of the cortex are not yet developed. The infant is not able to process the subtle details of the surrounding world.

For example, at the same spatial frequency, it will need more contrast than an adult to perceive the signal. It is more difficult for a baby to distinguish contrasts of distant objects because spatial frequency depends on distance. Imagine an approaching and receding lattice. At a far distance, it becomes frequent; at a near distance, it becomes rare — this is a high/low spatial frequency. The size (visual angle) of the image on the retina depends on the distance of the object from the eyes: the closer the object, the greater the visual angle, i.e., the spatial frequency decreases (cycles of light/shadow stripes within each degree of the visual angle). As the central nervous system develops, the normal ability for more accurate, detailed perception at a distance develops due to the improvement of integrating filters work, including encoding the spatial frequency. Throughout life, the accuracy of vision can deteriorate both due to the degradation of the eye's optical functions and integration functions in the central nervous system.

The accuracy of information processing also depends on the spatial arrangement of the fields of perception. Closer to the center of the retina (area of the central fossa), the sensitivity to light and the accuracy of perception are maximum. Primates have one fossa in each eye, but some birds of prey have two for more detail. Dogs and cats have not a fossa but a strip, and their vision detail is low. The matter is in the concentration of photoreceptors and the size of their field of perception. In the center the field is the smallest for registration of more detailed information. The distribution of zones of perception of different sizes throughout the entire field of view allows analyzing different spatial frequencies in a given part of the retina. It is like a sieve with a different grid size: large cell — low spatial frequency, small cell — high frequency.

But the accuracy of vision cannot be explained only by the topographic arrangement of receptors. Experiments with the visual system of flies, rabbits and monkeys have shown that the visual system can recognize displacements in the light signal that are less than the region of one receptor in the retina. Even a special term has appeared to describe such a phenomenon: hyperacuity. How is this possible? Are neurons "magical" and can transcend physical limits? No, they cannot. But they can do something else.

The authors of the book "Spikes: Exploring the neural code" explain: "Photoreceptors are usually small enough that they act as optical waveguides, and this effect can give a non-trivial (and wavelength-dependent) structure to the photoreceptors integration region. All of these effects can be summarized by an empirical angular sensitivity profile $M(\phi)$ — if cell number n in the array is stimulated by a point light source at angular position ϕ, then the photocurrent generated by the receptor is proportional to $M(\phi - \phi_n)$, where ϕ_n is the direction that this cell is "looking" in the visual field. If we know these angular sensitivity profiles, we can construct the response of an array of cells to a single spot stimulus and ask what happens when that stimulus moves by a small amount. What we see is that, even for an arbitrarily small displacements, something always changes — photoreceptors are not on/off switches, so they can respond by giving fractional changes in output when the stimulus is moved by fractions of a receptor spacing" (Rieke et al., 1999).

Unfortunately, the title of the book contradicts the description. To encode small displacements the outputs of a neuron cannot be spikes (identical discrete events). We are back to the Symphonic Neural Code hypothesis within TTT (see "Part Four. Algorithm of the Mind"). Action potentials are not purely digital on/off states. They are continuous oscillations with a specific waveform that can encode small and fast changes of a signal with sufficiently small and fast fractional changes of an impulse response of a neuron. The code should be sufficiently fast and accurate to be adequate to the encoded signals of the world. The angular sensitivity profile of the receptors that the authors describe is wave dependent in the sense that it encodes the incoming waves by the outgoing waves. The spiking neuron models (average firing rate and temporal code) cannot account for the actual hyperacuity of the brain. The output of a neuron must be meaningful even within one activity cycle, which means that the neural code is of a hybrid discrete-analog nature (symphonic code model).

In addition to spatial and temporal frequency filtering the system has to compress potentially infinite signal from the environment. The brain is limited by our cranium, and the Universe is not. There is no point in trying to "embrace the infinity." A more cunning approach is required: encoding the signal in a format and volume that is convenient and sufficient for adaptation. However, to define the compression level, the system needs to know what is important and unimportant. It has to compile a directory of what has already been processed. It is the reality model, which any individual living system receives genetically as a gift from ancestors, develops, and passes on to descendants.

Something the system can afford to compress, filter coarsely, even erase and ignore. The accumulated catalog of the reality model allows the system to operate in the "zombie" automaton mode in most cases. For example, we perform more than 90% of actions automatically and do not use higher cognitive functions that require a significant energy expenditure. Additional costs arise only when fixing errors or mastering a new skill.

If the environment remained unchanged, we would have turned into actual zombies. But all the time, it offers something new. It is necessary to differentiate and catalog this something: to sort it out by "shelves" and "folders," to give a unique index. In categorization, as a way to save space and time, humans have especially succeeded as a living system possessing abstract-verbal level of consciousness (more on that in "Part Seven. Inner Universe").

How did artificial signal processing technologies come to compressed formats (MP3, JPEG and others)? We studied the features of human perception of these signals using tests and determined what is important for the brain and what is not. It is one thing to use the deductive hypothesis that the finite cannot embrace the infinite to prove the principle of the system's operation. Another thing is to supplement it with evidence of an inductive property, specific psychophysiological indicators that can be established experimentally.

Tests show that some of the information is really not so important to the brain. Moreover, it is simply not able to process a part of the range of signal parameters. Thus, there is no sense in trying to pack too much into a code. Based on the tests, certain signal compression technologies have been developed to save computing resources, memory and energy.

The brain faces the same tasks. We can assume that it also uses compression technologies similar to artificial ones. If the physics of the signal processing process and the tasks of the systems coincide, then the hypothesis of the coincidence of approaches and technologies is quite working.

The main approaches on which the developers of artificial technologies base compression algorithms:

1. Initial redundancy of physical reality signals.
2. Only important signals need to be processed.
3. Identify important signals based on experiments.
4. Accept the loss of some information due to compression.

Obviously, the redundancy of potentially infinite signals for the finite brain is a hot topic for living signal processing technologies. The importance of a signal is determined based on previous experience. The system that considered any information not essential and made a mistake did not pass on its genes to the next one. Such an error in the game of Life is usually expensive. Experience is transmitted as a model of reality, changing in ontogeny and thus constantly dynamically developing in phylogeny. The system has to accept some information loss but must be careful with what to discard as unimportant. What are the main ingredients of compression technology based on the above approaches?

Hypothesis:

The brain "attacks" the problem of the potential infinity of environmental signals by using the information compression and indexing technologies combining the following operations: block splitting, sampling, quantization, and entropy coding depending on the probabilistic weight in the accumulated "catalog" of the reality model.

The brain organizes information in blocks at the earliest stage of the converting filters — for example, division of eye retina into rods and cones, center-periphery receptive fields. Then brain changes the base of the signal to "digest" it performing sampling and quantization. We can call it a kind of "dead zone quantizer" with thresholds for signal perception. It can also be described as the counter principle, when the signal is "counted" only when crossing the threshold, or the ladder principle when a signal with a specific parameter level can get to the next stage.

In mathematics, this is the principle of rounding to the nearest integer around 0.5. Imagine that you want to add non-integers. You have a set of numbers where the fractional part is either closer to zero or closer to one: 0.1236 or 0.953, etc. If you have few numbers, you can persist in adding everything up to all the decimal places. But if the amount of information is large, it is easier to round off those that are less than 0.5 to 0 and greater than 0.5 to 1. With a large number of digits, the result will not differ much from adding the original data. Yes, this is information compression, but strategically there is no significant loss, and the savings are significant.

The amount of information processed by the brain is hard to imagine. "If you are afraid of wolves, don't go to the forest." But the living system has "to go to the forest," that is, move in the environment. It can't do that without "tricks" in signal processing. You can walk in the forest in different ways. You can break through the thicket right through, or you can choose well-trodden paths. The direct route is not always the most efficient, neither when walking in the forest nor in processing and coding signals. The system has to employ wise coding methods.

One of the options is entropy coding: reducing the amount of data by averaging the probability of elements appearing in the encoded sequence. Morse code is an example of this approach in artificial technology. It is a simple code for transmitting a large amount of information over a limited communication channel (in this case, the telegraph). The principle is in assigning the shortest character (dot) for the most frequent letter in the alphabet (in English, it is E) and then increase the length of the code along with decreasing probability (increasing entropy). Thus, the code capacity and resource costs become balanced. The same principle of entropy coding is used in modern information technologies (Shannon-Fano code, Huffman code, etc.). Its essence: the greater the probability of a signal, the shorter the code sequence.

The same principle underlies the coding of environmental signals by the brain. If signals repeat frequently, it can use concise code to process them with a minimum investment of time, space and energy. The more predictable the data is, the more you can compress it. If a rare and unpredictable signal appears, then more resources will have to be spent on it. There are a lot of frequent signals, and less resources should be spent on each one separately. But the total resource cost will be large anyway. The brain minimizes costs by performing a "mapper" function. It selects patterns, encodes and reduces them to a model ("map"), which can be projected (superimposed) on incoming signals, registering only the difference between the bits and the bitmask.

Even in straightforward transmitter-receiver communication schemes, the receiver has an active role. The signal comes with distortions since any communication channel does not have an unlimited range with a linear function. Accordingly, adaptive reception and filtering schemes are required, as well as synchronization of both sides of the interaction.

In artificial technology, the receiver sends pilot tones to check the channel and set time markers. The transmitter sends training signal sequences for the receiver to test the channel and adapt its filters. The circuit is a feedback loop and is called "handshaking." Thus, self-tuning of the system takes place by using the existing model (bootstrapping) and its adaptation based on the current data. The analogy with the PAAL algorithm is obvious.

What is meant by an adaptive filter in artificial systems? "An adaptive filter is a system with a linear filter that has a transfer function controlled by variable parameters and a means to adjust those parameters according to an optimization algorithm. Adaptive filters are required for some applications because some parameters of the desired processing operation are not known in advance or are changing. The closed loop adaptive filter uses feedback in the form of an error signal to refine its transfer function" (Wikipedia "Adaptive filter").

Brain filters are adaptive in their physical and functional nature. They cannot be different since the environment parameters are dynamic, and their predictability is conditional and is a probabilistic category. The presence of a feedback loop to determine the difference error and refine the "transfer function" (creation of representations) is vital.

Also, the filters of any system that processes signals in real-time are causal, i.e., their impulse response (IR) depends on past and current inputs. The system must build causal chains, and this requires a point of reference. This point is the model of reality as a database of past signal values.

In artificial systems, filters are classified as finite impulse response (FIR) and infinite impulse response (IIR). FIRs work with a limited number of samples to achieve the output result. They are also called non-recursive due to the lack of feedback in the implementation scheme. They are applicable only in linear circuits, where the relative simplicity of their implementation and stability are used. However, in a self-learning system, it is necessary to use IIRs, which work with a potentially limitless set of input signals. Thanks to feedback, it can have an infinite impulse response.

Infinity in the name of such filters means that such a filter can feed itself with information, i.e., work in real-time. Thus, we again come to the lemniscate symbol as infinity, not in the sense of something unlimited in space and time, but as the iterative and recursive process.

Hypothesis:

The PAAL algorithm of the living Mind combines causality as taking data from the past and the present in real-time and the infinity of impulse response in the recursive feedforward-feedback loop, which allows the brain to work as a "time machine" to connect the past, present and future. The projected model of reality and introjected signals are convoluted for making assumptions about the state of the environment and actions in it.

Mathematically, the operation of the convolution of the past and present can be expressed as follows:

$$y[n] = \sum_{k=-\infty}^{\infty} x[k]h[n-k] = x[n] * h[n]$$

It is a symbolic representation of the signal processing by the filter, where y[n] are the output values, k is the signal, x[k] are the values of the signal sequence, h [n-k] is the time-shifted impulse response of the system (delayed by n). The algorithm performs the convolution of sequences x(n) as input values and h(n) as values in the past. The result is the integration of the signal shifted back in time and the current one, their internal product as an updated representation. This equation can be called the Mind's algorithm formula.

Convolution has the following properties:

1. Linearity — the possibility of product combination.
2. Commutability — the possibility to rearrange the sequence of filters' work.
3. Associativity — the possibility to sum a cascade of filters so that the impulse response of this system will be the result of a convolution of individual responses.

Brain filters are not only commutative but also mostly interchangeable. Associativity is the basis of what we call consciousness since it gives the impression of a whole product, while it is the result of the work of billions of components. The result of the filters' work is not just a sum, but synergy-convolution, as the superposition of the products of the individual work of the filters, and the product of their joint work in the past (model) on the product of their work in the present, in order to determine the future through acting, as testing the model. The integration of converted signals can be viewed as a convolution of different conversion products. The convolution theorem is applicable here — the Fourier transform of the convolution of two functions is the product of separate transformations and vice versa:

$$F\{f * g\} = F\{f\} \cdot F\{g\}$$

Signals in different modalities of perception from various amplitude-frequency ranges are converted separately. Then a convolution occurs when a general transformation product (reality model) is created. When the representations, as the characteristics of the incoming signals transformed and encoded by the neural network, are reproduced, the overall product is "laid out on the shelves" of individual products.

For the survival and adaptation of the system, the brain as a whole must be a stable filter with bounded input bounded output (BIBO), which prevents a computational "explosion" (pathology of the Mind) and a physical "explosion" (overload and substrate pathology). The Fundamental Stability Theorem from signal processing theory applies to the brain — a filter is BIBO only when its impulse response is summable, i.e., less than infinity:

$$\sum_n |h[n]| = L < \infty$$

The adaptive system's impulse response is summable: it restricts the incoming signals by its inner parameters and gives a limited output. In any other case, it is maladaptive, and its vitality is questionable. If flooded by the signals, the process may collapse due to a computational explosion, when the result at the output becomes non-summable, incoherent. Or the system sharply limits the input (dissociation), which in the long run means the same loss of adaptivity. No one is immune from overload, but if the filters themselves are not stable, then the risk increases even with the ordinary course of events.

The brain has ways to maintain stability. First of all, it is the coordinated and synchronized work of all elements of the system, which, in principle, ensures the integrity and coherence of the Mind. We will look at the physics, physiology, and technology of such a coherent state and its possible violations in further parts of the study. From the perspective of this part, we are interested in technologies using which the brain can cope with processing the signal flow.

We can call the brain a signal processing device with the statistical function to smooth out short-term fluctuations and identify significant patterns. By analogy with the algorithms used for this purpose in artificial systems, we can assume that the brain uses a moving average algorithm. In general, this is a function where each point value is equal to the average value for the previous period:

$$WWMA_t = \sum_{i=0}^{n-1} w_{t-i} \cdot p_{t-i}$$

where $WWMA_t$ is the value of the weighted moving average at point t; n is the number of values of the initial function for calculating the moving average; w_{t-i} — normalized weight (weighting coefficient) of the initial function t_i value; p_{t-i} — the value of the initial function at the moment, distant from the current one by i intervals. The function is calculated anew continuously, and the average moves by sliding along the time axis. The larger the set of values, the smoother the averaging result, but the greater the cost of processing and storing data.

The reality model of every living being is an accumulator of ancestors' experience and individual experience as an investment in the future. It is a "greedy snake" devouring its tail: the more signals, the more operations and storage capabilities it needs; the more possibilities it has, the more signals it wants to process. Memory, computing, and power resources develop, but they are not infinite. Processing stability requires self-limitation in accumulation. There must be a way out. We can assume that the brain uses a leaky integrator scheme. This means that some of the information is gradually leaked in the course of ongoing operations.

Hypothesis:

The brain works as a leaky integrator: the creation of a reality model is carried out with constant updating based on incoming data and the removal of part of the accumulated information to prevent possible overload of signal processing and memory systems. Thus, the brain maintains a balance of stability and dynamism of the model, working as a filter with an infinite impulse response in real-time. The PAAL algorithm's recursive nature of using the previous values as part of the input data allows the model to be adapted in an efficient mode of saving energy and memory resources.

Keeping the function smooth as a moving average that highlights significant patterns in signals requires a lot of input data. The projected model, as the accumulated volume, ensures the stability of the process, since the old version and the new version of the model have a small difference (the ratio tends to one):

$$\lambda = \frac{M}{M'} \approx 1$$

The more data have been accumulated, the less the impact of the change on the average. This is the "trick" of the algorithm's configuration, allowing it to work as a leaky integrator. On a large volume with a leak of a small fraction of data, the function will remain stable. It maintains high computing power at a relatively low cost. The closeness of a lambda to one is actually analogous to using a function on a large amount of data for averaging.

The following equation can describe a leaky integrator scheme:

$$y[n'] = \lambda y[n] + (1 - \lambda)x[n]$$

It shows that the old dataset with a partial leak (for example, at $\lambda = 0.98$, it is 2%) is filled with a similar amount of new information ($1 - 0.98 = 0.02$). If the lambda is close to one, then the brain creates an updated version of the model almost invisibly in terms of cost. The living system allows parts of the past to flow away to maintain stability in the present and future. Instability occurs when the model has a small base or a sharp change in signal patterns requires additional costs for updating the model and threatens with a computational "explosion."

Leakage and model updates should be small. Otherwise, the function will not be smooth, which carries a deadly risk of maladjustment. But the absence of changes ($\lambda \geq 1$) threatens a similar result due to the fixation and narrowing of the model and its possible detachment from reality. The brain in general tends to keep the ratio closer to one but must be prepared for sharp drops in its value due to increased uncertainty. The PAAL algorithm's smooth operation achieves a delicate balance between minimal information entropy and a potential surprise.

The formula shows a quantitative parameter, but there is also a qualitative question: how to determine what can be leaked and not throw out the "baby with the bathwater"? The self-learning principle of the PAAL as a comparison of projection and introjection counter flows corresponds to the task of revealing the significant difference and essential changes. But the algorithm is part of the answer. The system has to apply probabilistic approaches in forming a hierarchy of representations (more on that in "Part Seven. Inner Universe").

A critical factor in the stability of the entire process of environmental signals transduction is the reality model projection as an imposition of past data on the current input. For the system to function in principle, it needs this "lubrication": inserting a projection of a reality model into an iterative process, creating recursive ability of the algorithm to refer to itself and work in real-time without interruption. But to project the model, it must be saved to memory.

Chapter 5

Memory Technology

Questions of how the brain can a priori create its own goals and then find the appropriate search images in its memory banks are not well handled.

Walter Freeman III

The question posed by neuroscientist Walter Freeman in 2008 and cited as the epigraph to this chapter can be reformulated as follows: what is the technology and physics of the process of creating, storing and reproducing representations? This question is not well handled in the sense that it is not disclosed conceptually. The details of the physiological processes of the brain are quite accessible to the modern researcher. The fact that the question remains open only speaks of the failure of the theoretical modeling. There is no bridge between the phenomena of physiology and the explanation of the physics and technology of processes. The Teleological Transduction Theory (TTT) aims precisely at creating such bridges.

The expression "memory banks" used by Freeman reflects a long-existing notion that memory exists somewhere in special storage where the images are taken from when they are needed. But is there a place in the brain where all our memories rest and wait for us to pick them up? Neuroscience was actually looking for such a place for a long time. The logic was simple. The brain is anatomically divided into various areas that perform different functions, so we should be looking for memory as a special function in a specific place. This idea is supported by an enormous amount of data that shows the relation of mental functions with specific areas of the brain. On the other hand, the volume of data about dynamic distribution is no less impressive.

The struggle between localizationists and anti-localizationists has been going on for a long time. The paradox is that both positions are correct, but both are wrong. It is true that mental functions are associated with certain zones and that they are distributed throughout the system. The error lies in the concept of mental function. Both sides remain in a vicious circle of attempts to find correlates of mental functions in the substrate of the brain, forgetting that the concepts of these functions were formulated by psychology without any knowledge of the physics and technology of this substrate.

Ivan Pavlov described this dilemma as follows: "It is one thing if physiology, when analyzing life phenomena, uses information from other branches of knowledge that are more accurate than its own. And it is quite another matter when one has to borrow from a discipline, concerning which one must admit that it has not yet developed to the degree of an exact science ... Thus, a physiologist who has decided to study the activity of the cerebral hemispheres faces a dilemma: either to wait until psychology in its time adds up to an exact science, i.e., will designate its phenomena by the correct elements and develop their natural system. Only in this case could the physiologist successfully use psychological information to study the functions of the extremely complex structure of his object. Now I cannot imagine how it would be possible to superimpose the system of spaceless concepts of modern psychology on the material structure of the brain. Or another solution to the dilemma: the physiologist must try to go on a completely independent path from psychology, to find the basic mechanisms of higher nervous activity on his own ... And I dare to think that currently there are serious and positive reasons to accept that this way out is entirely normal and expedient, that its success is fully assured" (Pavlov, 1923).

Pavlov was right and wrong at the same time. Yes, psychological concepts are just phenomenological descriptions of the observable result of some internal process that psychology tried to define in terms that are not in any way related to the actual physics of the process. This made them non-scientific as they could not be empirically tested, confirmed, or refuted. But as history revealed, the completely independent path of the physiologists also turned into a phenomenological description of the physiology without an actual explanation of the mechanism. That is why his optimism about "fully assured success" was premature.

The founder of neuropsychology Alexander Luria, who all his life tried to combine psychology with physiology, admitted: "The whole history of attempts to localize mental processes in the cerebral cortex, whether these attempts were made from the standpoint of narrow "localization" or "anti-localization," retains one vicious position that causes the deepest dissatisfaction ... Genuine scientific analysis of the mechanisms by which the brain carries out an adequate reflection of reality was replaced by parallel statements about the "correspondence" of complex mental functions to limited or wide areas of the brain" (Luria, 1962).

It is exactly the essence of the problem: instead of looking for a physical and technological mechanism, neuroscience is trying to search for neural correlates of mental functions as defined by psychology. How can we find material correlates of something that is not defined physically? It is certainly a vicious position in all meanings of the word.

Almost a century after Pavlov's call and half a century after Luria's, modern neurophysiologists wrote: "Among physiological systems the brain in particular defies explanation, at least to a point where we can explain normal and abnormal behavior in mechanistic terms related to underlying cellular processes ... The question of how higher-level outputs (network activity, behaviors) are generated from lower-level properties (cells and synapses) has been considered for decades and still exists ... A complete library of individual components would not constitute understanding of a system ... For any complex system, there are many simple models we can invent to understand its behavior. The trick is to pick the right one. And that requires us to think carefully — to know something — about the essence of the real thing" (Parker, Srivastava, 2013).

The point is precisely in the correct model. But it is high time to move from the slogans about the necessity of such a model to its creation. The past hundred years have shown that the separate roads that psychology and physiology have taken led only to an impasse. Only building a bridge between these approaches is an entirely normal and expedient way where we may hope for success. This bridge is the physics and technology of the processes that are embodied in concrete physiology and lead to psychological phenomena. Throughout the study, we have been building bridges for many such phenomena, including the main one — the Mind. Now let's think carefully about the essence of the real thing called memory. It is certainly the core function. If the brain models reality, it must store this model, otherwise the whole process does not make sense.

Definitions of memory within psychology vary from very abstract classical ones like "a complex of cognitive abilities related to the accumulation of knowledge" or modern ones like "the faculty of the mind by which information is encoded, stored, and retrieved when needed." Whether we call it knowledge or information, whether we call it cognition or encoding, it does not get us out of the vicious circle between psychology and physiology if we do not elucidate the mechanism of information creation. There is no way to understand the technology of storing something if we do not understand what it is physically. To put it differently, if we do not know what the representations are, we cannot know how the system keeps them.

Here we must return to the basic hypothesis of TTT about the physical nature of representations as the hypothesis about the physics of memory will be directly related to it. To put it short, representations are wave patterns. Proceeding from this simple but concrete physical hypothesis, we can make assumptions about how these representations are stored. Moreover, the physical nature of waves suggests that there is no specific "memory bank." To illustrate this, we may use the holographic analogy again.

In the 1960s, when it finally became possible to use a stable and coherent reference wave in the form of a laser, the peculiarities of holographic memory drew the attention of engineers. Peter van Heerden of the Polaroid Research Center noted several advantages of this technology in terms of representation and storage (Van Heerden, 1963). First, it provided a very high recording density compared to traditional methods. Second, it offered high noise immunity and a good signal-to-noise ratio due to the distribution of information over the medium instead of narrow localization. The loss of a part does not lead to the loss of the whole, and even with severe damage, the carrier can retain information, albeit with a lower resolution. Third, it allows the entire medium to store a potentially massive number of individual patterns, not in sequence but in overlapping. This improves storage capacity, speed, and memory efficiency.

How is this possible? The secret is in the physical essence of the process: for each representation, you can use its own reference wave with specific frequency-phase parameters. The projection of such a reference wave initially creates a representation, which is saved on the medium in the form of a pattern unique to it. When reproduced, it chooses from the common resource by initiating the corresponding reference wave. This was not fundamentally new for wave technologies: radio and television broadcast different channels at different carrier frequencies, and for playback, it is necessary to switch the receiver from one to another. But in this case, it was not only about creation, transmission and reproduction, but also about saving. Here was the peculiarity, which drew the attention of van Heerden.

But here also lay the main problem. There was no appropriate material capable of such a feat of memory. Van Heerden noted the advantages in theory, but they had to be used in practice. Half a century has passed since then, but things haven't changed radically. Artificial memory technologies are developing forms that have already become familiar: magnetic, semiconductor, optical. But the holographic principle remains outside the mainstream.

There is a lot of research in industry giants and small projects. They periodically write about successes, but in practice, they do not reach large-scale applications. Some express the opinion that the idea is a dead-end and will soon become a historical curiosity. For people not connected with technology, holography is associated with color pictures with a semblance of a three-dimensional image. This view is very far from the holographic principle's real potential but reflects the state of our technology.

To understand the potential, it is enough to pay attention to the clear correspondence of the theoretical properties of holographic memory with the practical properties of brain memory: huge capacity, information density, a combination of differentiation and integration, the possibility for sequential and simultaneous processing, multilayer structure and associativity, speed, efficiency, resistance to carrier damage, flexibility, dynamism. Our memory works typically like an instant reader rather than a gradual scrolling of magnetic or optical media. In pathologies, there is a violation of the above basic properties, which manifests itself in a slowdown in extraction, disintegration, violation of the associative process, and even systemic decay (more on that in "Part eight. Dissonances of the Mind").

Our memory can be called "flash memory" for more reason than artificial media with this name. Modern flash drives offer advantages in compactness, strength, bulk, speed, and energy efficiency over others. But they are susceptible to external interference (electrostatic discharges), and they have a limited resource (accumulation of irreversible changes in the structure at each write-erase cycle). Writing, reading and erasing occurs in many cells simultaneously in relatively large blocks. The erase block size is always larger than the write block. This technology requires a high degree of identity for each cell. It becomes a problem when the media is reduced in size. Besides, when the write-erase cycles repeat, the charges of individual cells are gradually misaligned, and the writing or reading device cannot perceive them with high precision. The blurring of boundaries and the loss of identity leads to irreversible consequences for the quality of such memory.

A distinctive feature of such microcircuits is their strict hierarchy (blocks, sectors, pages). Any violation of the linear order during writing or reading will create an insurmountable obstacle to the normal process. Linearity also limits the simultaneity of parallel processes, so operations are performed using complex sequential algorithms. The presence of a certain number of defective cells and the algorithm's complexity create the need to equip microcircuits with a special command interface. A specific command is sent to transfer the memory page to a buffer, the data integrity is checked, if necessary, an attempt is made to restore and process further.

All of these properties create limitations in speed and efficiency. They have one reason stemming from the essence of the substrate itself: it is a passive environment in which a trace is imprinted as on "tabula rasa." It cannot adaptively change its parameters and correct the consistency of charges. A high degree of identity of passive information carriers becomes a paradoxical requirement: the cell must be a unique "personality" but also be a "slave." Slavery has never been effective for this very reason: lack of personal initiative and interest in labor results. The brain provides an example of a social structure: a democratic society of free individuals-neurons, connected into a single network but retaining personal adaptability and uniqueness. This creates the basis for operational efficiency, and a natural mechanism of synchronization provides coordination of interactions.

Our memory, of course, is not ideal, and even with a normal state of consciousness, it can show malfunctions. It also depends on the number of write-erase cycles and often accumulates irreversible changes. Our memory can overflow and freeze; it can reduce the storage capacity and lose the accumulated data. Memory is a physical process in a material medium, limited in its capabilities and resources. But it has a combination of properties that is not yet characteristic of any artificial media. They can show individual advantages in some parameters, but on the whole, they are seriously inferior.

Our memory has a "magical" mixture of distribution and unity, multi-layered nature and simultaneity, differentiation and integration, flexibility and reliability, speed and efficiency. Extraction of information can proceed in any way: sequentially, parallel, cross-wise (associative) and "jumping" (spontaneous transitions). The depth of the container seems inexhaustible with all the physical restrictions. It manifests itself both in the possibility of voluntarily extracting the required representation (thought, image, sensation, movement) and in the involuntary emergence of what seemed irrevocably gone. Our memory is very stable and dynamic at the same time. A truly systemic pathology is required for a substantial violation of the process to occur. A global stop occurs only with the complete destruction of the substrate (death).

It is impossible to "take out" a memory block and stop its work because there is no such block. There is no "hard drive" or "flash drive" to provide local storage of "files." Memory is a system function and a system process. This formulation is too general and not new at all. To say that memory is spread and not local simply states a fact that has been known for many decades. A physical explanation of all these "magical" properties of memory is required. We need an answer to the question: what process can provide such a system function?

Back to the holographic analogy again. The holographic principle's theoretical possibilities, which are still awaiting their practical implementation, are based on wave processes and their synchronization. The fact that we cannot yet use them to the full in artificial technologies does not mean that they have not already been used in live ones. But on what basis can one conclude that living systems use them? Even if we did not know anything about oscillatory processes in the brain, we could make a simple deductive hypothesis. If there are coincidences in the properties and manifestations of external physical phenomena and internal ones, then perhaps they have fundamentally the same physics and technology. There may be differences in details, in the substrate, in the frequency range, and so on, but the principle of the universal mechanism remains. In addition to this speculative assumption, there is enough empirical evidence at this stage of our knowledge of the nervous system to draw an inductive conclusion that these observations coincide with such a deductive hypothesis.

In holographic technology, the frequency-phase coupling (synchronization) of the object and reference waves creates the prerequisites for those properties that favorably highlight the theoretical potential of such memory. But the coordination mechanism is not enough. A mechanism for creating information and storing it is required. Any analogy with modern computers falls apart at this stage because, in them, the principles of storage and retrieval are based on sequential, bit by bit, reading. With all the enormous speeds of modern processors, whose frequencies are orders of magnitude higher than those of the brain, this technology does not provide all the properties characteristic of our memory.

Here is an opinion expressed quite a long time ago by the authors of one review article: "It has been observed that the time required to execute computer instructions is in the order of nanoseconds, whereas neurons take tens of milliseconds to fire ... Thus, it is argued, the brain must operate quite differently from computers ... If one had to search through one's memory serially, the way conventional computers do, the complexity would overwhelm any machine. Thus, the knowledge that people have must be stored and retrieved differently from the way conventional computers do it ... Unlike digital circuits, brain circuits must tolerate noise arising from spontaneous neural activity. Moreover, they must tolerate a moderate degree of damage without failing completely ... Human memory representations, and perhaps many other cognitive skills as well, are distributed spatially, rather than being neurally localized. This appears to contrast with conventional computers, where hierarchical-style control keeps the crucial decisions highly localized ... In computers, memory is static: once an entry is put in a given location, it just sits there until it is operated upon by the CPU... (In the brain), representation itself is doing something — there is no separate processor working over it ... Conventional rule-based systems depict cognition as "all-or-none." But cognitive skills appear to be characterized by various kinds of continuities ... Conventional models are dictated by current technical features of computers and take little or no account of the facts of neuroscience ... The fact that this gap between high-level systems and brain architecture is so large might be an indication that these models are on the wrong track ... Perhaps this includes even the assumption that the description of mental processes at the cognitive level can be divorced from the description of their physical realization. At a minimum, by building our models to take account of what is known about neural structures we may reduce the risk of being misled by metaphors based on contemporary computer architectures ... The question is whether there is anything to be gained by designing "brain style" models that are uncommitted about how the models map onto brains" (Fodor, Pylyshyn, 1998).

Unfortunately, the situation has not changed fundamentally since then. Conventional models are still on the wrong track. They think in terms of the "all-or-nothing" discrete neural code paradigm. That is why they fail to account for the observed speed of the brain functioning despite neurons being so slow in comparison to computer processors. They fail to explain the observed properties of natural memory technology: high capacity, flexibility, efficiency, fault and noise tolerance. They seem to ignore the physical reality and do not care about how the models map onto it.

The inertia of old paradigms is very strong. But our brain is capable of building a map of the world, so it must be capable of mapping itself. We need to acknowledge the dead-ends even if it means understanding that a lot of time and recourses were wasted. This is the first and hard step. The next one will be easier. We will just have to look into the continuities of the physical processes.

The secret of the brain is in the wave nature of the encoding mechanism. Waves form, store, and read information not sequentially bit by bit, but from all participants in a specific wave pattern simultaneously. Complex information can be generated, written, and read within one clock cycle, determined by the carrier wave's base frequency, without the need to accumulate and read the successive bits along a linear chain.

Neurons are active self-oscillating systems with subtle mechanisms for creating the parameters of their impulse response in the general continuum of the phase portrait of the system. There is no "grandmother neuron," but neurons participate in ensembles to create various representations of grandmother. There is no "picture of grandmother" stored in a certain place. There are settings of numerous neurons that participate in representations that we call memories of our grandmother. One neuron may be part of the wave producing a memory of our grandmother's pie, another — in a memory of her face when she smiled and yet another — in a memory of her face when she cried.

The brain does not have separate recording devices, information carriers and reading devices. Each neuron in every part of the brain performs all these functions. The process of information creation, storage and reproduction is carried out in our heads by the same substrate. There is, of course, specialization determined by the role of this particular neuron or population in the overall technological chain of the PAAL algorithm. But in general, our model of reality as saved representations is at the same time unified and distributed.

The "depository" of our memory is so deep because there is no depository. It sounds paradoxical, but if you look at the physics of the process, the paradox disappears. Representation as a wave of oscillatory activity of neurons participating in a given pattern is determined simultaneously by each element's settings and the entire ensemble. In such a process, each neuron is important, but it can also be missed. A wave should not involve every individual oscillation, but if there is no critical mass of oscillators capable of supporting it, there is no wave.

In modern artificial technologies, the principle of content-addressable memory works for fast search: the algorithm looks throughout the database and provides answer options (as opposed to random access memory, where you have to input the search address). In the holographic principle, such associativity is initially present because even a tiny fragment of a recording can be used to reconstruct a complete interference pattern. The larger the piece, the better the resolution (more certainty), but a less certain result is often sufficient.

Instant natural associativity of our memory is manifested in the fact that sometimes we do not understand how one thought or image "clings" to a whole string. These can be not only thoughts but also sensory-motor representations. The sound of familiar music can cause an entire flurry of sensations, emotions, memories, images.

We can create an associative array by an effort of will, but spontaneous branching is also possible. This property of our memory in psychology is usually explained by a moving "beam of attention." This creates the illusion of explanation, but it remains a mystery what this "beam" is physically, not metaphorically. In light of the TTT hypotheses, it becomes material and really analogous to a beam of a light wave: the movement of the projection of the reference wave within a given pattern initiates other patterns in which populations or individual neurons of a given ensemble take part. Frequency-phase coupling generates a cascade of different melodies, harmonies, and rhythms of the Mind. It resembles a musical medley, a mixture of motives united by one theme, placed on the same rhythmic meter, written in the same key or different keys, but connected by a modulating chord sequence.

The reference wave of projection can synchronize "adjacent" representations purposefully or give spontaneous and unexpected variations on the theme. It remains for us to narrow the result down to significant and relevant representations, if we were in an arbitrary search, or to surrender to the free wandering of such a wave in the hope of "reaping" in this field of the reality model some exciting and new solutions. An associative "potpourri" can become not just a superimposition of old motives but turn into a new piece of music.

Attention to the problem of associative choice in artificial memory systems based on the holographic principle was drawn back in the 1960s. The idea was that for the stability of the search process, a scheme with two different memory regions is required: in one, all holograms are stored together, which provides a fast search over a small fragment, but gives mixed results; in the other, information is accumulated separately for accurate retrieval. But this division creates additional resource requirements. Is there a need for this? If we are talking about an artificial memory system, then the problem of choice is facing the user. He has the right to turn to different regions: first, determine the set, and then select the appropriate result by narrowing the search window.

In a living system, memory and user are fused. As a user of its own resource, the system can at any time make a choice and stop the search. Of course, the problem of choice exists, but the system has accumulated experience to separate the "wheat from the chaff." It can continue the search if it needs and has time. There is no requirement for division into blocks of memory. Moreover, the problem of choice turns into an advantage of choice. The expansion of associative search is a vital adaptive advantage that has arisen due to the development of higher integrators, which allow forming and maintaining a wide range of representations. One of the signs of the systemic disintegration of consciousness is the violation of the associative function resulting from the desynchronization of the integrators (more on that in "Part Eight. Dissonances of the Mind").

Since different oscillators can participate in the same wave, and various waves can use the same elements, the physics of the wave process and synchronization provides the basis for dynamics and variability. The natural associativity of such a physical process ensures the evolution of the variability of thought processes. Thus, the physiology of creativity is based on the physics of wave phenomena. Any new thoughts (discoveries, inventions, creations) do not arise from scratch but are associative branches and variations of the interference pattern of superposition and synchronization of representations accumulated in the reality model as wave patterns. A narrowing of the associativity of thinking indicates a violation of the interaction and synchronization of wave patterns, i.e., it is a symptom of the collapse of the normal process of consciousness, pathology and maladaptive state.

In the evolution of the Mind, the development of the structures of the nervous system that ensure the creation and accumulation of representations (higher integrators) led to an increase in the ability for variable associative branching. The development of the brain substrate aimed to provide a wide range of reality model and formed the basis for creative problem-solving. This ability is especially expressed in humans due to a major development of higher integrators of the cerebral cortex.

To illustrate the associative creative process, we can again take a musical analogy since it simultaneously shows the essence of technology and is based on the same physics of oscillatory processes. For a musician to create something original and new, she must have a developed base of accumulated "bricks" that make up the temple of music. Rudiments of rhythmic, melodic and harmonic structures are necessary for the performance of old pieces and the creation of new ones. When a musician can combine these rudiments in harmony, she can create a harmonious variation as an innovation and a rudiment for future innovations. What appears to be rudimentary today was once the creative enlightenment of past generations. What seems to be revolutionary is an evolutionary transition to a new level.

Despite the enormous potential of the holographic principle, artificial technologies face the problem of a substrate for keeping information. This problem arose in our attempt to develop external resources in addition to our internal ones. At previous stages of evolution, living systems had the same storage problem, but they had no choice: they had to use what they had, i.e., themselves.

What are the requirements for a substrate that could provide this potential? It must have nonlinear properties to adaptively change its characteristics under the influence of waves passing through it. Modern artificial media change their state under the influence of signals (this is the essence of their function), but their adaptability is minimal. Usually, it comes down to recording discrete states, which are then transformed into a continuous form by other system blocks.

In a living multicellular system, each cell is simultaneously a recording, transforming, saving, reproducing device, and part of the general distribution medium. It is initially adaptive and nonlinear. The ensemble of such cells becomes an even more nonlinear dynamic and complex system. It is an advantage inherent in the very process of unification. As specialized cells that create a model of reality and ensure the movement of the whole organism on its basis, the nervous system is an active nonlinear medium in terms of its composition. Each cell adaptively changes the parameters of its impulse response to wave patterns created by ensembles in which it participates.

Memory is not an object stored in a container, but a wave process in an oscillatory medium. Not only is the creation of a representation initially a transition from the world of signals to the world of the internal code, but each reproduction of a wave pattern means the activation of oscillators, the characteristics of which could change. Each time we reproduce the memory anew. We do not get it from the "storage shelves" of our "memory banks" unchanged.

The number and composition of oscillators participating in the wave of representation can vary. The same elements can be involved in different acts of memory as evoked wavefronts, and the same memory can result from the work of a changed set. This explains the simultaneous dynamism and stability of memory as a function of storing, evoking and reproducing representations. The wave nature of the process also explains the associativity of memory work: one wavefront can become a trigger for another since neural oscillations involved in it can begin to interact and synchronize with other oscillators.

The system stores the settings of the network elements to create and reproduce a wave with a specific shape (parameters). These settings can change quickly if a signal triggers large shifts in an impulse response of filters. That is why emotionally salient and action-related events leave more profound traces in the memory. They just involve sufficient flows of neurotransmitters to influence the settings. For other signals, repetition is required. Popular wisdom says: repetition is the mother of learning. Specific electrochemical processes leading to a change in the conformation of neuron receptors and a general restructuring of intracellular processes are behind this fact of life. But consolidation occurs both with repetition and with strong one-time exposure to neurotransmitters. To put it simply, we remember what we repeat, what is interesting to us, and what we accompany with actions. These are the three "faces" of the mother of learning.

Please note, that the valence of the emotion or action does not matter. In this sense, the mortal danger is interesting to us and our actions to cope with it will leave deep traces in our memory. The secret is again in the chemicals involved. For example, neurons of the hippocampus have a high concentration of glucocorticoid receptors. Glucocorticoids (for example, cortisol and corticosterone) perform the function of mobilizing and activating the body. Their level in the blood rises sharply during shock conditions. They are active mediators in higher integrators, including the hippocampus. And this is entirely logical: if the hippocampus is an integrator and participates in the formation of representations, then in a situation of danger, it must be activated to keep up with the rapid changes. In experiments where the activity of glucocorticoids was blocked, the subjects had difficulties reproducing emotionally significant events. With an increased level of cortisol, better memory consolidation is observed.

But there is the other side to the coin. Too high a concentration of excitatory neurotransmitters leads to excitotoxicity or even the death of cells. That is why there is also an effect that has long puzzled psychology: people do not remember highly traumatic events or have distorted memories of them. There is a longstanding explanation in psychology that it is a result of the workings of a psychological defense mechanism. But the truth of the matter is that it's the workings of the physics and technology of memory. Of course, it can be interpreted as defense, meaning that the system defends itself from overexposure. On the other hand, if all living systems did not remember traumatic events, their defense against the dangers of life would have been zero.

Let's summarize our description of memory physics and technology.

Hypothesis:

Brain memory is the process of encoding, storing and decoding representations as wave patterns where many elements of the neuronal system can participate at once as oscillators with specific characteristics tuned to the parameters of this particular pattern. The system needs to store only the impulse response settings of these oscillators. The rest is done by wave physics. The main properties of memory created by wave technology are the combination of high capacity and precision, differentiation and integration, sequential and parallel processing, layering and associativity, stability and flexibility, speed and efficiency, dynamism and fault tolerance.

Memory works according to the PAAL algorithm where reference waves of stored representations are projected and meet with introjected object waves to form an updated version of the reality model. Memory has the same levels as the Mind, starting from basic sensory-motor representations and up to the highest abstract-verbal ones. Though every neuron contains a memory function, populations of neurons have their specialization as filters of the algorithmic technological chain. Thus, we can say that memory is localized and distributed at the same time. This physiological fact is explained by the technology and physics of the process.

The only difference between the above definition of memory and the standard ones that we have quoted earlier in the chapter is that it adds a hypothesis about the physical mechanism and a technological chain. If we take this definition out of the context it may seem a "magical" explanation as all complex details of how the physics of waves provide for the astonishing properties of our memory are not disclosed. But we had covered a long way to this point in our study, starting from fundamental physics, going to biophysics, and moving to the physics of the brain in the previous volumes. Thus, we have a sufficient level of understanding of the details to be able to formulate hypotheses on such a general level and not leave them "in the air" without any support. Moreover, there is still a lot to explain in the further chapters of this volume and the next parts of the study, and it all concerns wave processes.

Let's add some more details that will show that the explanations within TTT that stem from physics can serve as predictions of the observed physiological and mental phenomena. As the above hypothesis states, wave physics provides for a set of features that seem incompatible with each other, but nevertheless, each of them is vital. For example, stability and flexibility. The mainstream paradigm says: memories are encoded by modification of neural wiring. Of course, if we think of the neural activity as electrical impulses going through wires, we are bound to model memory as establishing the connections to produce the encoded patterns of spikes. Next, we inevitably think of representations as spike trains going along the wires. Logically we assume that for the representation to change a rewiring has to happen. We know from physiological data that the actual rewiring is a long process. But then we find out that representations and neural populations that encode them are actually fast-changing phenomena. We have a physiological fact that changing connection takes longer than the observed speed of change of the ensembles representing a signal. We are at an impasse. Our logic was flawless, but we started with the wrong assumption. We spent a lot of time and energy within the paradigm but it looks like it was a waste. We are frustrated.

To deliver the emotional impression let's take an excerpt from a press article honestly called "Neuroscientists have discovered a phenomenon that they can't explain." Here is how the author and the neuroscientists that he interviewed describe the situation: "Carl Schoonover and Andrew Fink are confused. As neuroscientists, they know that the brain must be flexible but not too flexible. It must rewire itself in the face of new experiences, but must also consistently represent the features of the external world. How? The relatively simple explanation found in neuroscience textbooks is that specific groups of neurons reliably fire when their owner smells a rose, sees a sunset, or hears a bell. These representations — these patterns of neural firing — presumably stay the same from one moment to the next. But as Schoonover, Fink, and others have found, they sometimes don't. They change — and to a confusing and unexpected extent … (Researchers) recorded the activity of neurons in the rodents' piriform cortex — a brain region involved in identifying smells. At a given moment, each odor caused a distinctive group of neurons in this region to fire. But as time went on, the makeup of these groups slowly changed. Some neurons stopped responding to the smells; others started … The same phenomenon, called representational drift, occurs in a variety of brain regions … Its existence is clear; everything else is a mystery. Schoonover and Fink told me that they don't know why it happens, what it means, how the brain copes, or how much of the brain behaves in this way. How can animals possibly make any lasting sense of the world if their neural responses to that world are constantly in flux? If such flux is common, "there must be mechanisms in the brain that are undiscovered and even unimagined that allow it to keep up," Schoonover said. "Scientists are meant to know what's going on, but in this particular case, we are deeply confused. We expect it to take many years to iron out" (Yong, 2021).

Of course, if we think that a representation is a train of spikes of a certain neuronal network connected by long-term wiring, we expect them to reliably fire when their owner smells a rose or hears a bell. Then we find that neuron ensembles taking part in the same representation of a smell or a sound can change a lot faster than the rewiring can take place. We are at a dead-end. We are deeply confused. Should we change our initial assumption about the nature of the code and the physics of the process? It seems logical that if we have facts that do not fit the model, we should change the model. It is not so easy but it is the only way out of an impasse.

The researchers emotionally and logically react to the unexpected representational drift: "How fast does it go? How far does it get? And … how bad is it?" How does the brain know what the nose is smelling or what the eyes are seeing, if the neural responses to smells and sights are continuously changing? … Schoonover and Fink compare the discovery of representational drift with the work of the astronomer Vera Rubin. In the 1970s, Rubin and her colleague Kent Ford noticed that some galaxies were spinning in unexpected ways that seemed to violate Newton's laws of motion … Drift indicates "that there's something else going on under the hood, and we don't know what that is yet," Schoonover said" (Ibid).

We have devoted the first two volumes of the "Symphony of Matter and Mind" series to the physical issues, including the explanation of phenomena on the macrocosmic level based on the mechanism "under the hood" proposed within the Theory of Energy Harmony. Here we will just point out that the analogy is perfect: neither the galactical drift nor the neuronal drift is explained by the current mainstream models. The only thing that standard physical models could offer in connection with new discoveries was the concepts of "dark matter" and "dark energy." Such names just highlighted the dark spots in the models. Neuroscience has its dark spots too. Representational drift is not the only or best-known problem, but it came as a total surprise. That is why the reaction of the researchers expressed in the interviews is so emotional (they usually hide it in the dry reports on the experiments). But it is good that neuroscientists try to avoid references to "dark forces" when they encounter something not yet explained and look for a rational explanation.

But perhaps the explanations for galactic and neural drift are not as far apart as our brains and galaxies. Are we witnessing the same mechanism under the hood of the cosmic and neuronal galaxies? Should we get emotional about finding a solution to those mysteries? The answer to both questions is Yes. Our emotional systems are the evaluating module in the brain. If we get confused and frustrated, it means we were wrong. If we get excited, we may be on the right track. If we take all the baggage of knowledge accumulated throughout this study the mechanism does not seem so unimagined. We may even find it rather banal in hindsight.

Our excitement may fade but we do not feel frustration. On the contrary, each time new data fits into the model, we feel satisfied. How do we reach that state? By developing our models of reality, including our models of ourselves and our brains. They need to be stable and flexible at the same time. If a model is too flexible it is not a model but chaos. If a model is too stable it is not a model of reality but a dogma detached from reality. The only way out for a living system is to keep the reality model in a balanced state by letting the projected model interact with introjection and updating it in case of discrepancies. We have to be open to the new.

Here is how the article ends: "So now what? "There's a real hunger in the field for new ideas," Fink told me, "People are really desperate for theories. The field is so immature conceptually that we're still at the point of collecting factlets, and we're not really in a position to rule anything out." Neuroscience's own representations of the brain still have plenty of room to drift" (Ibid).

The accumulated evidence that contradicts mainstream theories of neural coding is so massive that one would think they should just evaporate under this pressure. But we cannot substitute something for nothing. Though the field is desperate for theories, neuroscientific publishers are not eager to provide their space for new ideas. Paradigms do not give up easily. Meanwhile, practical neuroscientists work in the absence of a coherent model of a mechanism under the hood. They are just collecting facts that do not fit into the assumptions of the mainstream models and feel deeply confused. So now what? Let's try to make a representational drift away from the mainstream. After all, it is not so difficult if we are not constrained by the need to "walk the line" for financial and political reasons.

Here are the questions asked by Schoonover and Fink about the drift: why it happens, what it means, how fast it goes, how far it gets, how the brain copes, and how bad it is. If we look at the Symphonic Neural Code (SNC) hypothesis and other hypotheses that we have proposed in this and previous part of the study we can answer the questions and not wait many years to "iron it out." As we have covered the issues in detail elsewhere, we will just give short answers here.

Why does it happen and what does it mean? Interestingly, the author intuitively used the keyword — flux. Representations are waves of oscillatory activity of neurons participating in a given wave pattern. Each oscillator is important but the same pattern can be formed by different oscillators. For the pattern to be stable (the same smell of a rose or the sound of a bell), there has to be a critical mass of co-tuned oscillations. This connection is not about wiring but about frequency and phase coupling. The anatomical connections are also important but they are part of the same wave physics (more on that in the next chapter). Waves and their synchronization provide stability and flexibility at the same time as different oscillators can participate in the same wave, and various waves can use the same elements. The representation is stable but not the participating neurons.

How fast and how far does it go? The representation dynamics are within a timescale that corresponds to the dynamics of the world. The intraneural impulse response properties that determine the participation of a specific neuron in a wave pattern can change much faster than the interneural connections. The fact is that the actual rewiring happens rarely and is a lot slower than the flux of the Mind. On the other hand, impulse response properties are stable and do not change if there is no need for an update. Change of the ensemble does not mean changing the instrument or its part. Different ensembles can play the same music of the Mind representing the smell of a rose or the sound of a bell.

Representational drift is the result of a universal wave mechanism. Thus, it can go as far as any neuron in any part of the brain. The difference is in the functional role of this particular neuron and population. Schoonover and Fink studied the piriform cortex (Schoonover et al., 2021). They call it the sensory hub. Within TTT technological brain map it is considered a primary filter-integrator of the olfactory system that forms the representations of the current introjected signals. Due to this functional role, its dynamics are fast. But the higher filters of the neocortex to which it sends the data for multimodal integration are the projectors of the reality model, so their representational dynamics are more stable. The same goes for other modalities of perception and corresponding areas. Any primary and intermediate integrator, including subcortical ones, is more prone to representational drift than the higher one. By the way, this explains our susceptibility to dogmatization of representations at the abstract-verbal level of consciousness. A lot of time must pass and a lot of data from reality must accumulate in order for the model to undergo cardinal changes.

The studies of drift are only in their infancy but there is sufficient data confirming its existence in the hippocampus, parietal sensory-motor cortex, and occipital visual cortex (Ziv et al., 2013; Rubin et al., 2015; Driscoll et al., 2017; Marks and Goard, 2021; Deitch et al., 2021). On the other hand, studies show long-lived (from weeks to months) neural ensembles dynamics responsible for producing stable motor skills and behavior (Dhawale et al., 2017; Katlowitz et al., 2018). Variation of timescales across brain areas and in different trials confirms the TTT prediction that the speed of representational drift is a reflection of the functional role of a specific integrator.

How does the brain cope? The question is ill-posed as it proceeds from the assumption that the drift is some hindrance that should be coped with. We should get back to the first question about the role of the drift and its mechanism. The answer is in the physics of waves and their interactions that solves the paradox of stability and flexibility of the same system. The drift is part of the mechanism. The confusion disappears once we understand the mechanism under the hood.

How bad is it? First, as drift is a part of the natural mechanism that the brain uses to create representations and the reality model in general, it cannot be bad. Second, the facts can be bad only for a model that cannot fit them. In this sense, the drift is very bad for neuroscience textbooks that the article mentions. Moreover, it is bad for the whole field of theoretical neuroscience that has been comfortably flowing in the mainstream. Though the explanations proposed in this chapter are physically grounded and empirically confirmed, they may take many years to "iron out" because they offer new streams that seem dangerous. But it is only by expanding our conceptual comfort zone that we can get out of the quagmire of confusion that makes us feel so uncomfortable when confronted with a reality that contradicts our notions.

Let's check the predictions of the memory physics and technology hypothesis by the reality of clinical cases of disrupted memory function. We will start with the Henry Molaison (H.M.) case. The uniqueness of this case is that the physiological origin of the pathology is beyond doubt and there is a detailed description of psychological manifestations. It is, probably, the most famous clinical case concerning memory. In 1953, patient Henry Molaison had large parts of the medial temporal lobe (MTL) of both hemispheres removed during surgery in an attempt to cure epilepsy. It has receded, but he acquired a major memory pathology. He was diagnosed with anterograde amnesia (inability to form new memories).

Reproduction of previously stored representations was mostly intact. Molaison could evoke even early childhood memories. He did not create new declarative (semantic and episodic) memories but retained the ability for procedural memory. He could master a new motor skill but did not remember how he acquired it. The ability for spatial memory and orientation was at a normal level. He could reproduce the layout of the residence to which he moved five years after the surgery, which means that his brain could still produce some types of new representations and keep them. His working memory functioned well. In general, Molaison was able to remember explicit information over short periods of time. For example, he did well on word retrieval tasks but failed at the level of language comprehension and production at the sentence level.

Physiologically MTL includes hippocampus and parahippocampal, perirhinal, entorhinal cortex zones. The hippocampus (from Ancient Greek ἱππόκαμπος — seahorse) "sits" between the deep subcortical structures and the higher structures of the cortex. Its communication with the cortex is carried out through the entorhinal zone, which itself is connected by countercurrents with other cortical and subcortical structures. Only these physiological facts lead to an obvious idea that this region is central to some function.

According to the TTT, new representations are settings of brain filters' impulse response that allows them to produce wave patterns. The hippocampus plays a key role in this process as the creator of new wave patterns, to which higher integrators of the cortex will subsequently be tuned. Based on this hypothesis about the function of this area of the brain, clinical observations of its lesions or other causes of malfunction, indicating a variety of pathological manifestations, become understandable. But historically, it is the diversity of manifestations that has led to confusion about the function.

There are three main versions of the hippocampus function: purposeful behavior, spatial map and memory. As the authors of one review write, the challenge remains to understand how these functions are related and correctly note that "the seemingly disparate functions of the hippocampus need not be mutually exclusive" (Ito, Lee, 2016). From the TTT perspective, these are manifestations of one function of the hippocampus, which stems from its place in the technological chain. Let's see if the new hypothesis can shed light on the old data.

The version about the role of the hippocampus in purposeful behavior was especially popular in the 1960s. It was based on the observation that with hippocampus lesions, animals became hyperactive, could not inhibit behavior, poorly learned to avoid danger, and hardly coped with the situation of conflicting signals. Is this version wrong? Not really. It just speaks about everything in general and nothing in particular. If a part of the brain performs some vital function in forming a model of reality, it is sure to have a role in purposeful behavior. After all, the Mind in general exists to ensure such behavior.

Suppose we approach the hippocampus as an integrating filter and its function as participation in the formation of new representations. In that case, all aspects of the symptomatology when this element of the system is affected can be explained. The difficulties in learning, coping with ambiguity and controlling behavior are the result of the problems with forming representations and updating the reality model. How can the living system assess the situation and behave adequately if it does not have an idea about current signals?

The version about the role of the hippocampus in spatial navigation was based on the very long-standing idea of the existence of "cognitive maps" in the brain. The victory was celebrated when the authors of the discovery of "place neurons" were awarded the Nobel Prize in 2014. This is how the committee described the discovery: "This year's Nobel Laureates have discovered a positioning system, an "inner GPS" in the brain that makes it possible to orient ourselves in space, demonstrating a cellular basis for higher cognitive function … The discoveries of John O'Keefe, May-Britt Moser and Edvard Moser have solved a problem that has occupied philosophers and scientists for centuries — how does the brain create a map of the space surrounding us and how can we navigate our way through a complex environment?" (Nobel Committee Press Release, 2014).

John O'Keefe has been looking for a "cognitive map" in the brain since the 1960s. He implanted electrodes into the brains of rats and recorded the electrical activity of individual neurons or small groups of neurons adjacent to the electrode (O'Keefe, Dostrovsky, 1971). He found that there are neurons in the hippocampus that fire when the rat runs through a specific place in the maze and when it returns there. This fits into the idea of localizationism: if a neuron is triggered when the system does something in some place, then it is a "place neuron." The correlation between the signal and the activity of neurons was taken for a direct causal relationship.

May-Britt and Edward Moser repeated his experiment decades later. They studied the activity of neurons in the hippocampus and the entorhinal cortex, located in the temporal lobe and associated with the hippocampus (Fyhn, Molden, Witter, Moser, Moser, 2004). There they found other elements of the "inner map" that also reacted to the geographic location of the animal and called them "grid neurons." These neurons did not activate just at one point in space, but in many different ones, organized in a hexagonal structure, similar to the one that cell towers of mobile communications form. When the rat ran in one direction, the cell fired at regular intervals at the grid points. Sequential signals of the "grid neuron" mark the distance and the sequence of firing of several such cells — the direction of movement. Later "head direction neurons" and "boundary neurons" were discovered in the same cortex in rats, bats, primates, and humans (Zhang et al., 2013).

Pay attention to the fact that at first, "place neurons" were found in one area of the brain. The researchers decided that they solved the riddle of the cognitive map. More than thirty years passed, and another zone was identified where neurons are activated by the same stimuli, but in a slightly different way. They decided that now they had found the map for sure. Even the Nobel Committee decided it was time to announce the discovery of the navigator in the brain and awarded all participants. By the way, the analogy with GPS is characteristic: we inevitably look for analogies in the work of the brain with the work of artificial systems that perform the same function — signal processing. But if you look closely at the research of the laureate authors, there was no hint in it about the mechanisms of this processing. Let's read the authors' report carefully in order to understand the actual result of the study.

The authors confidently start by stating the presence of specialized cells but eventually admit that work and interaction mechanisms are unknown. The beginning of the article: "The mammalian space circuit is known to contain several functionally specialized cell types, such as place cells in the hippocampus and grid cells, head-direction cells and border cells in the medial entorhinal cortex (MEC). The interaction between the entorhinal and hippocampal spatial representations is poorly understood, however … The mechanism that generated the space signal, and its location in the brain, were not apparent" (Ibid).

The authors used optogenetic labeling, where the introduction of a special virus creates a targeted expression of a light-sensitive transgene in neurons. It allows using the activity caused by a light pulse as a tool for tracking direct neural connections. What did they find? "A key finding of the optogenetics experiment is the identification of border cells with direct inputs to the hippocampus" (Ibid). Okay, but the fact that neurons connect has been known for over a hundred years. The details of the links are refined all the time, but what does this give for the stated purpose of searching for a mechanism? End of the article: "Whether place cells can be formed only from this subset of the population, and whether the properties of such place cells match those of observed cells, remains to be determined ... Addressing the detailed mechanisms of place field formation would require genetic tagging at the level of individual cells — a technology that is on the horizon, after the emergence of single-cell monosynaptic tracing technologies as well as methods for intracellular recording and stimulation in MEC cells of behaving animals" (Ibid).

Any word about the mechanism of the formation of representations? At the very beginning of the article, the authors wrote that the mechanism was not apparent. The verb is used in the past tense as a hint that everything became clear after their research. But no, at the very end, it becomes clear that nothing has been clarified: the study of detailed mechanisms requires further detailed study, and the question remains open. As a result of all the subtle and lengthy investigations, the mechanism of the formation of representations remains poorly understood. The question of whether the work of the cells under study corresponds to the function "place neuron" invented for them by researchers is unanswered.

In short, the study did not provide answers to the questions posed by the researchers, and the hypothesis was not confirmed. The authors believe that this is due to insufficient data. Perhaps the problem lies elsewhere. First, the problem is that, according to the paradigm of localizationism, the search for the location of the mechanism continues, while the mechanism is not an object sitting in some place but a technological process. Secondly, the main problem is that there is no definition of what we are looking for. What mechanism should be located somewhere? What "beast" are we looking for in a dark room? We must first decide what a "cat" is before looking for it in either a dark or a light room. The authors not only failed to find the location of the mechanism in the brain but also did not give any version of the mechanism that generated the space representation.

Perhaps the reason for the lack of an answer to the question of how the representation of a place is created is not in the lack of detailed study, as the authors suggest, but in the incorrectness of the question posed? After all, the question "How?" is a purely technological issue. You can persistently investigate the details of a device, but without insight into what a given device does, it is impossible to understand how it does it. Perhaps the question is missing: what is the technological function of the device being studied? The brain is not only biological tissues made of cells but a high-tech device. Cells and tissues have specific functions to achieve the tasks facing them, as specialists within the framework of the goals of the whole organism. It sounds like a trite statement, and it is unlikely that the authors disagree with it. But the simplicity of such a statement is deceiving because it requires specific questions and answers. And here lies the secret of the authors' failure in answering the question that they asked themselves.

The question was too general and did not provide directions for finding an answer. Instead of specifying the question, they preferred to concretize the data they received. Physiological detailing and even future monosynaptic tracking at the level of individual cells and intracellular recording, which the authors rely on, will only provide an accumulation of data about the process. But a process model will not arise by itself if we do not pose the right questions to these data. To illustrate this let's take an analogy with GPS used by the Nobel Committee.

Imagine that we do not know anything about the technology of this system. Before us is just a body that determines its position on the ground in an incomprehensible way. There may be a suspicion that it contains some internal organs, tissues, cells, i.e., elements that perform specific actions and provide the process of finding the location of the entire device. We can look inside the body and see that there are specific elements and chains that connect them. At this stage of cognition, we may have an idea: in some zone or even in a specific part, a function "lives." And it's true: parts perform some function. The question is: which one? And here, a subtle but decisive moment arises. If we assume that the function of these parts is to determine the location of the device, we are making a category mistake: we confuse the concepts of "purpose" and "function." It is not as innocent as it might seem.

People often confuse these concepts. What difference does it make to say "the purpose of the device is to find a place" or "the function of the device is to find a place"? When we use this device, there is no difference. For us, as users, purpose and function become synonymous. But if we want to understand how it works, we have to be very careful with the words. Otherwise, we can lead ourselves into a dead-end. The description of the target system function is not a description of the function of the system elements. If we do not ask ourselves what the components do, then we will not be able to answer the question of how they do it.

We know that the purpose of a GPS device is to find a location, but what does it do to find it? It receives a signal from the environment (in this case, signals from satellites), converts it along the ADC chain, creates representations of the received signals, overlays them on the existing map of the area and gives the final picture of comparison of projection and introjection in the form of its own position on the ground. Signal processing and creation of representations is a function of the entire technological chain of system elements. In this general function, they have specializations as filters at a particular stage of the process.

To answer how they do it, we have to understand first what do they do. In artificial technologies, we know this from the beginning since we are their creators (direct engineering). In the case of living technological devices, we do reverse engineering, trying to understand how the existing device works. We need insight into the function. Only then all the knowledge we have already accumulated about the processes and any further detailing will contribute to the solution of the problem and not lie in a poorly understood and not clear heap of puzzle pieces, instead of forming a coherent and clear picture.

The question of function is primary, the question of technology is secondary. In neuroscience, researchers often try to answer the question of technology (what is the mechanism) without answering the question of function (what does the mechanism do). Unsurprisingly, the answer is poorly understood and unclear all the time. Why has this been the case for decades with a persistence worthy of a better application? The problem is that there is an illusion of understanding the function, and it is very stable, difficult to correct, despite the apparent discrepancy between assumptions and results. This illusion arose because general psychological descriptions of systemic processes were taken for the functions of system elements. The researchers are trying to apply old terms to new knowledge about internal processes' details and find correlations of these concepts in the brain. Many people point out this mistake, but the inertia of the illusion is powerful.

The authors of one review article wrote: "For many years, researchers tended to associate preexisting verbal terms, such as memory, planning, envisioning the future, volition, and decision-making, with different and distinct brain structures. Oftentimes, even studying the same structure or system generated seemingly contradictory hypotheses depending on the chosen approach, preconception, or experimental method used. A striking example is the hippocampal system. Lesion data in humans provided evidence that bilateral removal of the hippocampi produced severe and irreversible amnesia. Single unit studies in the hippocampus and entorhinal cortex of animals gave rise to the prominent theory that the fundamental function of the hippocampal system is supporting spatial navigation, assisted by inputs from the head-direction system. In contrast, studying the collective behavior of hippocampal neurons by recording the local field potential (LFP) offered the conclusion that hippocampal theta oscillations are an unmistakable reflection of voluntary action. These independent ideas of hippocampal functions persisted in parallel for decades without true interactions, apart from occasional polite gestures and references to the competing frameworks" (Buzsáki, Peyrache, Kubie, 2014).

There are many "contrasting" results, which at first glance do not intersect with the role of the hippocampus in spatial orientation. For example, the research shows that the hippocampus is involved in memory tasks in various modalities (Wood et al., 2000; Sakurai, 1994; Deadwyler et al., 1996), temporal discrimination tasks (MacDonald et al., 2011; Nakazono et al., 2015) or even in the processing of abstract information such as rules (Nakazono et al., 2019). From time to time, one of the competing concepts received prizes at the highest level. It seemed like an achievement, a victory. But there was no overall victory in the form of an answer to the question of how the system works. Perhaps the secret is that the function itself is mislabeled? If we proceed from the physics and technology of signal processing and the creation of representations, we understand that the function of a given element of the chain is determined by its place in the general algorithm and technology. This insight sheds light onto the old data that cease to contradict each other but confirm the same functional role of a zone.

Within the TTT hypothesis about the levels of representations and filter architecture of the brain that we considered in previous parts of the study, the hippocampus is considered as a part of integrators circuit that is responsible for representations of higher level which we usually call conscious. From this perspective, all seemingly contradictory results of experiments that show its participation in various activities of an organism and all aspects of the symptomatology when this element of the system is affected can be explained.

Here we get back to the main versions of the role of the hippocampus: purposeful behavior, spatial map, and memory. As we have already noted, there is no way that a living system can have purposeful behavior if it cannot form representations of current signals. And, of course, new representations include the spatial aspect of the environment. The hippocampus is not a "cognitive map" but a part of the filter architecture that creates representations of the various signals of the environment. No wonder, its activity is registered when the system is involved in different cognitive tasks.

That is why Molaison's brain could create representations at an implicit level, but completely lost the ability to create conscious representations. He could orient, but could not explain to himself or others how he did it. His internal GPS functioned well enough. When the Nobel Committee wrote about "solving a problem that has occupied philosophers and scientists for centuries," it missed this widely and long-known fact that disproves the hypothesis for which it issued the prize. Moreover, as we have seen, the authors themselves admitted that they did not reveal the mechanism for creating representations of a place, and, therefore, did not solve the problem. However, the expectations were so great that even the absence of a solution was taken as a discovery.

Let's take another famous clinical case that is directly related to memory. In 1985, a musician and musicologist, Clive Wearing contracted encephalitis which damaged the medial temporal lobe and the prefrontal cortex of both hemispheres of his brain. This is what physiology says. Psychology says that he has anterograde and retrograde amnesia. Technologically speaking, it means that his brain cannot form new representations and has lost settings for the old ones. Life for Wearing narrowed down to a moment. Every minute is new as his memory does not store anything for more than 30 seconds. In a conversation, he can answer a question, but he cannot keep the conversation flowing for the obvious reason: he does not remember what was said moments ago.

But he can keep new sensory information. For example, after watching a movie, he does not remember ever seeing it, but when seeing it again he is able to anticipate certain parts of the content without remembering how he learned them. Despite having no memory of specific musical pieces when they are mentioned by name, Wearing remains capable of playing complex piano and organ pieces, sight-reading, and conducting a choir. Every time his wife visits him, he makes acquaintance with her but emotionally he knows that she is the woman he loves. He says he doesn't have any thoughts or dreams.

He was asked to keep a diary in the hope that it would help him to remember something. But there are only repeated inscriptions in it: "Now I am really, completely awake. Now I am perfectly, overwhelmingly awake. Now I am superlatively, actually awake." He writes, and then after a while, he looks at the record again, does not believe that he wrote it, crosses it out, and writes again that this time he has really woken up. He's been making these records for many years without any change in principle.

This clinical picture suggests that the higher integrators of the brain are affected, which are responsible for the creation and projection of a conscious picture of the world. Physiologically, the difference between Molaison and Wearing is that the latter has lost not only MTL, but also the prefrontal cortex. Psychologically the difference is that the former at least has past, and the latter does not have anything except seconds of the present. It is a condition that is hard to imagine to the full extent. The patient himself describes it as death.

According to the hypothesis within the TTT, storage of representations means saving settings for the impulse response of neuronal populations to certain wave patterns. In Wearing case, the consolidation of new settings does not happen, and the old ones disappeared. Waves of explicit memory do not occur because there are no oscillators to create them. However, the preservation of the implicit procedural memory indicates that the sensory-motor integrators are intact and, in general, the system still creates representations of the control of movements and the state of the body. Moreover, the presence of short-term explicit working memory indicates that this function is the prerogative of primary and intermediate integrators.

This clinical case also highlights once again that in order for representations to get into long-term memory, they must pass through a short-term memory which is very limited in time and space. It clears up very quickly to maintain the speed of processing incoming information. This means that the information from the primary signal processing enters it and, as it fills up, is transferred further along the chain to the long-term memory. If long-term memory does not function, we can call the result as living here and now literally. There is only the present, but there is neither the past as an accumulated model of reality, nor the future as its projection.

Let's expand a little bit on the importance of working memory and how it functions. For the primary processing of information, fast and efficient working memory is required. A living system's ability to create representations of signals with the necessary and sufficient level of accuracy and volume depends on the depth of this memory. But it should not be too deep; otherwise, speed and efficiency will be lost. Technologically the working memory of the brain works as a circular buffer with tail-end recursion. This gives recursion the form of a simple iteration, allowing it to avoid the stupor of endless self-call. The working memory prefers to quickly grasp what it finds and sort it out. But it also needs depth: it creates the possibility of accumulating information in long-term memory. The more information it processes, the more will be saved in the system.

Long evolution has taken us to the current level of working memory depth which is estimated as 7 ± 2 elements. Our closest relatives on the evolution branch, chimpanzees, have a working memory of two elements. Numerous experiments and observations of chimpanzees in the natural environment have shown that their working memory is limited to two concepts in manipulating non-living objects and in social interaction. This is evidenced by their behavior when using tools, by the length of combinations of gestures and sounds, and by the results of experiments to test the direct memorization of symbols (Read, 2008).

For example, from our point of view, such a simple action as using one stone to smash a nut against another stone (only three objects) is not a trivial task at all for chimpanzees. Data collected over 16 years of observation of wild chimpanzees have shown that only 1/4 of the population learns this simple method. And if a chimpanzee has not learned until 3-5 years old, it will never learn. The process is gradual: up to two years, manipulating one object, after two years — two items, after three years — three. It would seem such a small step, but it is so hard that a very small percentage of the population makes it. The secret is in the depth of recursion required to combine the sequence into an associative chain. A difference of only five elements of working memory distinguishes man from ape. The evolution of technologies as complex means of labor and culture as the ability to create associatively constructed means of communication depends on it.

We can judge what the world was like for our ancestors by the preserved relic cultures. We can take the example of the Piraha tribe who live almost without contact with the rest of civilization in the depths of the Amazon jungle. The recursive depth of their memory is well manifested in their mathematical apparatus. We can say that it is absent. Experiments have shown that they have a poor understanding of numbers greater than three. In general, they do not have numbers, but only the concept of few/many. They use three words to describe the quantity, roughly referring to the relative difference, which is not fixed and depends on the situation.

Scientists who studied Piraha did the following experiments: they laid out spools of thread up to 10 pieces, and Piraha had to lay out the same number of balls. If the spools were in a straight line, then there were no errors. If they were covered with a screen, or the balls had to be set perpendicularly, then errors immediately began. It means that they did not count objects, but simply related them.

The small recursive depth of working memory does not allow them to create deep long-term explicit memory. The Piraha tribe is called the happiest tribe on Earth, because they live literally in the here and now. They possess the immediate reward reality model and have no interest in the past and the future. Missionaries failed to convert Piraha to Christianity for a simple reason: they are not interested in the story of Jesus, if he lived a long time ago and the narrator does not know him personally. For them, a "long time ago" is a non-existent concept. They trust only their feelings or the direct experience of another person. The tribe does not know its history, and the world for them is constant and cyclical. They lose all interest in an object if it disappears from view and interaction. Even having made some reserves in order to exchange them for some item they need, they use this item only here and now. They can easily throw a shovel into the water after using it to dig a hole.

The evolution of our working memory in phylogeny is reflected in its evolution in ontogeny of each individual. Children under one year do not have enough working memory depth for a recursive function. Research shows that our children acquire sufficient depth only by the age of five. Which, by the way, allows them to begin to learn the grammar of speech and complex sentences. In general, our working memory's depth creates a breadth of range for our model of reality which is stored in the long-term memory. But, as we saw in the case of Wearing, having a working memory without a long-term memory does not mean a happy life in the here and now, but a state of disability and the agony of the "forever today" (the title of his wife's book).

Let's consider one more example of a brain pathology leading to total impairment of memory function and disability — Alzheimer's disease. Here is a cross-section of the brain of a normal older person and a patient with the diagnosis:

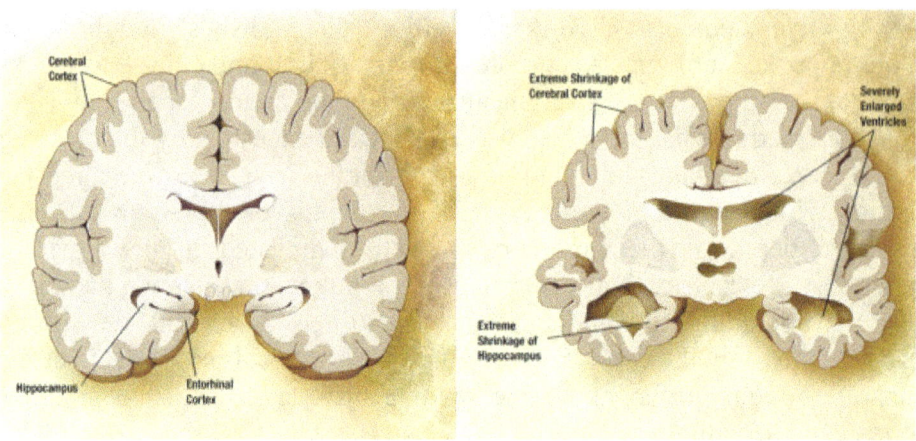

We can see an almost complete disappearance of the hippocampus and entorhinal cortex accompanied by a considerable shrinkage of other brain tissues. This is the physiological side of the picture. What is psychological? At the early onset, a person does not recollect events that happened just recently. But older episodic memories are intact and semantic memory is quite robust except for difficulties in finding words for immediate use. Implicit procedural memory is also not impaired, and a person can still cope with life on his own.

Usually, relatives and friends describe these symptoms as if the person starts to forget things. But he did not forget them; he just does not remember them. For him, they never happened as his brain just did not register them. It is no surprise, that other people find it hard to convince this person that the events actually happened. His brain does not create new representations. A person can remember past events before the disease onset, but he just does not have any memory of the new events.

This subtle difference between "forgetting" and "not remembering" can be explained only by our memory physics and technology. According to the TTT hypothesis, memory is the process of creating, storing, and reproducing representations as wave patterns. Thus, remembering the signal means that its representation has been encoded, consolidated, and stored in the settings of the system's elements. If the change in the settings did not take place, then there is no consolidation of a new representation. We do not remember the event. It just did not exist for us from the start. We did not forget it. Forgetting is when the stored settings fade for this or that reason. The event existed for us but ceases to exist.

The result of not remembering or forgetting is the same, but the technological road is different. Forgetting can happen due to the initial weakness of the settings change. It could be the outcome of a not sufficient number of elements involved in representation, as the more oscillators take part in a wave pattern, the more saturated the structure. Subjectively, we experience it as a more detailed, vivid memory. Otherwise, it is just not stable enough. The number of elements can be initially small or can diminish as a result of their pathology or death. But fading of memory could be just a part of a normal reconsolidation process that changes the settings, allowing to clear limited storage resources for an updated version of the reality model. In general, we all remember things and forget them for different reasons that are all reflected in the technology of our memory.

In Alzheimer's disease, the degenerative process continues to spread. When it reaches the higher integrators of the cortex all cognitive activity is disrupted and a person becomes fully dependent on caregivers. Now it is not only that new memories are not created, but old ones are erased. Even close relatives become strangers. The person loses the past and consequently the future. Planning, abstract thinking, executive function, attention, speech, and goal-directed motor activity — all fade due to the deterioration of the substrate of the brain that is responsible for encoding, storing, and decoding representations of the signals of the world.

Chapter 6

Neural Code Transmission Technology

The anatomical synaptic connection is necessary but not sufficient for transmission to occur.

Walter Freeman III

We come to the question of the physical nature of the substrate that ensures the functioning of the wave technologies of the brain. We need to remember that waves are the propagation of oscillating energy. Waves can move and spread in any direction, expand and contract, have any individual shape and create various patterns of interaction. They can have a tiny area of occurrence but spread almost instantly. The wave speed depends on the parameters of the medium: the less inert the medium, the more elastic (active, dynamic) it is, the higher the rate. Our nervous system can hardly be called inert: in it, every element, down to the intracellular levels, is an active oscillator tuned to interact.

Wave propagation, its shape (parameters), its localization in the space of a given environment depends on the dynamics of the elements participating in it. Thus, a wave of neural activity is not propagation of vibrations in all directions with the same properties, but a structure with a purposeful movement and the meaning of its properties for network participants creating, transmitting and receiving representations as inner meanings for the whole system. When we talk about brain waves, we should not imagine them as excitations walking back and forth from edge to edge and embracing everything that "comes across" along the way.

The brain is a technological device that uses the laws of physical processes for a specific purpose: the formation of the reality model as information patterns. These patterns have to be not only created but transmitted, stored and reproduced in a coherent manner. Wave processes have substantial technological potential, but they also have limitations. Technology is the art of making the most of different opportunities while minimizing constraints and even turning disadvantages into advantages.

In a passive and homogeneous medium, a wave from a source propagates in all directions. On the one hand, this is advantageous since it allows information to be transmitted to any receiver. But on the other hand, it is a disadvantage since the brain has to target its messages. The transmitter and receiver must be connected, speak the same language and know each other. The transmitter does not just send out to all "radio stations" but sends address messages. The receiver does not catch everything in the "air" but is tuned to a specific range and code.

An ideal emitter can theoretically transmit waves in any direction with the same intensity. It is an advantage. But for a real transmitter to approach the ideal one, the medium must allow such propagation: it must have conditions for the formation of waves of any kind (traveling, standing, transverse, longitudinal, mixed, solitons, spherical, flat, spiral and others). The medium itself must approach the ideal. There is no ideal, but you can strive for it. For this, the distribution medium must be active, dynamic and adaptive. Thus, it becomes part of the system that links the transmitter and receiver.

Any natural medium leads to energy dissipation, and the wave intensity is inversely proportional to the distance, i.e., decreases with its increase. To overcome this disadvantage, the environment must provide not only targeted transmission but also the maintenance of waves. Such a medium should be an active conductor for a complex spectrum with different phase and group velocities. It must deliver wave packets quickly and accurately to the address, preserving their structure and content.

Anisotropic medium has the property of wave scattering (dispersion). But this disadvantage can be an advantage since it means the dependence of the wave speed on its frequency, and such a dependence allows differentiating waves. The spectrum becomes not a general background but concrete "figures." Wave packets can have the ability to overlap each other without changing their structure (the principle of superposition), i.e., spread in the environment as if they were "transparent" to each other. But, if necessary, wave patterns must overlap each other to create synchronization patterns that connect different representations. Differentiation (analysis) must be combined with generalization (synthesis). Representations must be unique in structure in order to encode specific signals but also merge into a coherent and unified model of reality. What environment is capable of providing such a combination of different and even somewhat opposite requirements?

The leading neuroscience paradigm says that neural code transmission is about discrete impulses sent over electrical wires of neuronal connections. It is a technologically simple idea, but it contradicts the reality of the speed and efficiency of information flows in the brain. Moreover, it leads to conceptual dead-ends as such a technology is not able to provide the above requirements. Within the TTT framework, we explore the idea that the "relay race" of the neural code is a wave process and that the brain is an active and elastic medium that allows rapid and accurate wave propagation while preserving frequency and phase characteristics in which information is encoded. For this, it has to be a set of waveguides. The propagation of waves in such an environment has unique properties that are very suitable for solving technological problems facing a living system.

Let's take again artificial technology for an analogy. Waveguides (electromagnetic, acoustic, optical, and others) have a special configuration to guide the movement of the wave. Usually, they are made in the form of shielded tubes, inside of which there is an environment with special parameters for the propagation of waves of a specific frequency range. As a result of these properties, the waveguide is a speedy means of transmitting energy-information while maintaining the specified wave parameters with particular frequency, amplitude and phase characteristics, as well as with a certain set of wave modes (harmonic oscillations).

Each mode is a wave structure with a carrier frequency. A set of vibrations' frequencies constitute a spectrum as a superposition (combination) of various oscillations with a fixed phase relationship. Modes can coexist and move independently of each other in the sense that they retain their individual frequency and phase structure. Due to the limiting conditions of the waveguide as a channel, there is a set of frequencies and waveforms that can propagate through it. There are cutoff frequencies, and the lowest one will be the carrier, the fundamental frequency, the base for superimposing the entire harmonic spectrum of the waveguide. In essence, we are talking about the structure of a conducting medium with given parameters to create an effective information transmission channel as a set of synchronized wave structures inside the waveguide. A synchronization pattern is created that has certain characteristics and structure, which can be transmitted with minimal distortion and as efficiently as possible.

Hypothesis:

The nervous system, as a medium for wave propagation, is an active waveguide. Such an environment can provide the following prerequisites:

1. Reducing the effect of dissipation of wave energy (function of active amplification).

2. Differentiation of wave packets due to dispersion (dependence of speed on frequency) and the creation of frequency cutoff modes for propagating waves with a specific amplitude-frequency characteristic (filter function).

3. Limiting the wave propagation area (purposefulness, targeting).

4. High and adjustable speed of information propagation in the form of wave patterns with a stable structure.

5. Combining different wave types (traveling, standing, transverse, longitudinal, mixed, solitons, spherical, plane, spiral and others) of various phase and group velocities (even with opposite direction).

6. Integrating different wave patterns (representations) into a general synchronization picture (reality model) while maintaining a unique identity of a potentially infinite set of such patterns.

The nervous system is simultaneously a transducer, transmitter, repeater, active waveguide, receiver, modulator and integrator. The settings of the network elements can change and thus change the waveform. Waves can use different sets of cells or activate the same groups in a different sequence. This means that they can coexist in the same anatomical space but be in different parameter spaces and, accordingly, create various representations of a potentially infinite space of parameters of environmental signals. Does this hypothesis about the physics and technology of transmission correspond to the observed physiology?

The brain is mainly composed of white matter (myelinated axons):

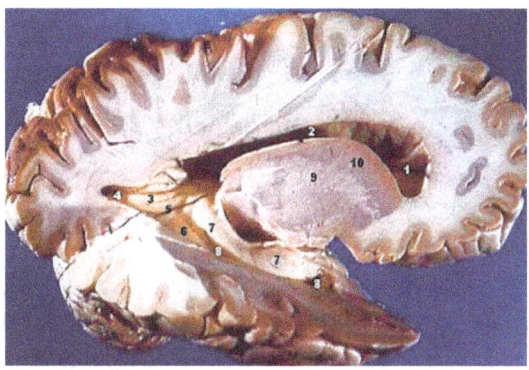

The same is true for the periphery. Spinal cord section:

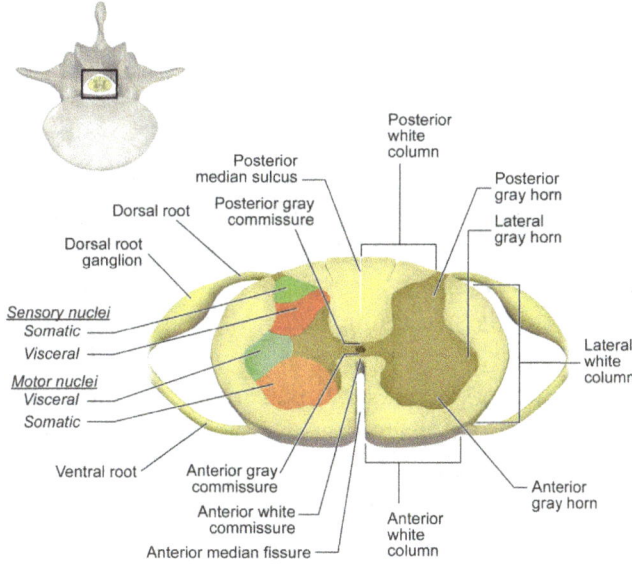

The Wikipedia article states: "Long thought to be passive tissue, white matter affects learning and brain functions, modulating the distribution of action potentials, acting as a relay and coordinating communication between different brain regions" (Wikipedia "White matter").

Does it explain at least approximately how this active role is carried out? No. The "Functions" section consists of five sentences, the meaning of which is that "white matter is the tissue through which messages pass between different areas of grey matter within the central nervous system" (Ibid). That is all about the substance which makes up the bulk of the brain. But where is the active role of this substance in the description? The author, probably, would be happy to disclose it at least within the scope of the article in the encyclopedia. But there is no information and no models of how this active role is fulfilled. There is not even a discussion of versions.

If we proceed from the hypothesis that discrete spikes transmit information along the wires of the network, then the axons (white matter) become "the tissue through which the messages pass." It's just a bunch of wires, and that's where their role ends. Then we run into dead-ends of the standard models of neuroscience. But what is the active role of white matter in influencing brain function? It is the same as any conductive medium, and it only appears to be a passive conductor. In fact, this is the environment where the wave patterns of representations form, spread, synchronize, disintegrate, and again form. Our thoughts, images, sensations, feelings, movements, everything that we call the Mind (Consciousness, Soul, Psyche) "walk, run, fly" in white matter.

The morphology of axons is not as simple as one would assume based on the hypothesis that these are wires for transmitting linearly propagating signals. To begin with, they are not flat, one-dimensional structures. But such classical models of spike propagation as the Hodgkin-Huxley model, the disadvantages of which we have already considered, speak of one-dimensional spike propagation. Further models refined the details and hypothesized the nonlinear properties of the process dynamics, but they still reduced the axon to a one-dimensional idealized "wire." It's not so much about trying to simplify but about ignoring the wave-like nature of the process.

But the reality is a little more complicated. Axons are three-dimensional, and their cross-sections are very far from symmetry in three-dimensional space (they are not circular). The length-to-diameter ratio strongly affects the parameters of the processes, especially for the areas of the cortex, which have relatively short axons. The volumetric nature of the propagation medium and the parameters of this volume, the shape of the channel radically affect the propagation of waves. The paradox is that these parameters are ignored by the leading models of neural communication in neuroscience. Why is such obvious empirical evidence excluded from models? The answer lies in the initial hypotheses of these models that say nothing about wave processes.

Axons have a complex shape and internal structure. But why did the evolutionary process need such a complication? If you look from the point of view of the transmission of electrical signals through wires, then there is no answer to this question except the helpless line: nature has contrived something for reasons that are poorly understood. But nature, or rather, the process of evolution as an increase in the efficiency of performing tasks, does not complicate without necessity. It is not nature that is wrong, but the hypothesis of the mainstream. The task in a neural network is not to transmit similar spikes with minimal losses. The fundamental point is that it is not just energy that is transmitted, but information as encoded patterns of energy. It seems that this is an obvious idea, and theories of neuroscience keep saying about information. But when it comes to physics and physiology, it turns out that the models are about the transfer of energy, not information. If we proceed from the hypothesis that complex wave patterns of energy in which information is encoded are created and spread, it becomes evident that nature has not been mistaken and has not "overdone" with the morphology of its structures.

Let's take one study that looked at the correlation between the structure of axons and the parameters of the activity of globular bushy cells (GBC) located in the ventral cochlear nucleus of the sound localization pathway in the mammalian auditory system. This is a good model for analysis as this circuitry processes interaural time differences (ITD) and interaural level differences (ILDs) with exquisite precision. The authors reported: "GBC-myelinated axons, and in particular those of cells tuned to low sound frequencies (for which ILDs are minimal, and processing ITDs of only microseconds is important), deviate significantly from the canonically assumed structure in a paradoxical way: low-frequency fibres are thicker but exhibit a shorter internodal length than high-frequency fibres. Our simulations as well as our recordings in vitro and in vivo indicate that this makes AP conduction particularly fast and precise … Differences of axon diameter and internode length may tune different length branches of the same myelinated axons to provide the required AP arrival times" (Ford et al., 2015).

The authors do not speak about waves but their results are obviously about oscillatory and wave processes. The parameters of axons are tuned to the frequency parameters of the transmitted signals precisely for the reason that they are not just signals anymore but information. The structure of the wave is not only about arrival time but about all the complexity of amplitude-frequency characteristics and phase portrait. The neurons work hard to create the intricate details of their oscillations, which encode the details of the oscillations of the incoming signals. Sending them across the wires that will not keep the information intact does not make sense. This is why the morphology of the axons is so complex down to the smallest details.

For example, there is an intracellular structure involved in forming the information transmission medium: neurofilaments. These are protein polymers with diameters of nanometers and micrometers in length. Together with microtubules and other elements, they form the cytoskeleton of the neuronal cell. It is believed that their primary function is to create the structural shape of the axon and regulate its diameter in order to influence the conduction speed. Only the diameter factor is considered as a parameter of the wire for power transmission. As an additional function, they are credited with the role of axon transport (movement along the axon of various biological materials — mitochondria, vesicles, signaling molecules, proteins, structural components of the cytoskeleton, ion channels). It seems like an internal "freight railroad" for the delivery of the neuron structure elements in general and the axon in particular. But each neurofilament is a spirally twisted thread, i.e., this "road" is not at all straight.

It is also interesting that the number of these elements in axons increases with the development of the brain, and their distribution becomes more diffuse. There is another "strange" pattern: as the network develops, the shape of the axon section moves further and further away from the circle and becomes very uneven. What does that mean? Why does development lead to a deterioration in the ideal shape of the "wire"? From the point of view of classical theories of neural communication, this is nonsense. Therefore, such a phenomenon is simply ignored, excluded from models, formulas for the electrical conductivity of neurons and calculations.

But if we look at the axon not as a simple power line but as a medium for the propagation of wave patterns, then, firstly, some points become clearer, and secondly, we no longer need to ignore the facts to preserve the familiar model. If the axon is an active medium for propagating wave patterns, then its entire complex and inhomogeneous structure acquire the meaning of modulation of the flows of these waves. Even an irregular cross-sectional shape is essential since it helps regulate the vectors of ion fluxes, giving them a specific shape and direction. In addition to the unevenness of the "walls" of the waveguide, there is also the inhomogeneity of its internal environment with various organelles and other elements. From this perspective, the role of neurofilaments can be much more interesting than adjusting the wire diameter.

The propagating wave has a complex frequency and phase structure and must be transmitted quickly and accurately. The shape of flows may cause the spiral shape of neurofilaments. But most likely cyclical causation is at work: the shape of the elements depends on the function, and the performance of the function depends on the shape. In this sense, other "oddities" of axons can also become understandable, such as the myelin sheath's twisted shape with many layers. When interpreting the function of an axon as an electrical wire, the mystery is the form of insulation itself and the seemingly excessive number of myelin layers.

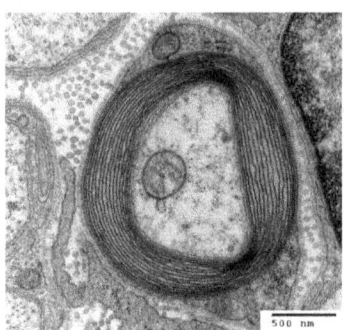

Why does nature need such a complex structure if the task is simply to transmit an electrical signal through a wire? For an analogy, let us recall the history of the evolution of information transmission at a distance. The first telephone lines were single-wire laid openly. It is crucial that these lines conveyed not just a flow of energy, but information, a structured flow. But what happened as a result of this wiring solution? There were many other electrical lines around, causing noise on the telephone line and disrupting the structure of the message. The power lines grew, and so did the noise. Various engineering solutions have emerged to protect information from interference.

For example, Alexander Bell, in 1881, invented a twisted pair, and its advantage was so obvious that by 1900 the entire American telephone network was redesigned based on this idea. What's the idea? The degree of connectivity of conductors in pairs increases and interference affects all wires equally, reducing external interference and mutual interference. Information is transmitted at different frequencies using an inexpensive and efficient twisted pair circuit. Without this scheme, it is challenging to transmit both low-frequency signals (telephone "beep") and high-frequency signals with many information streams simultaneously. It was implemented in ADSL technology, which replaced the modem connection. Interestingly, the dense twisted pair is also physically more resistant to wear and tear and is more compact. Twisted pair is now a standard solution for structured cabling systems in communication technologies as a physical medium for transmitting the information. Due to its low cost and efficiency, it is used as the most common solution for local area networks.

There are other solutions: fiber-optic networks that transmit light waves over glass and plastic wires over long distances quickly and reliably; wireless networks using different frequency bands and the air as a carrier of waves. They all have their advantages and disadvantages. Several solutions are often combined to balance the benefits and minimize the drawbacks.

Let's return to the communication system within a living organism. Evolution has used different solutions for communicating information. Many living organisms do not have a nervous system at all and use other methods of internal communication. Many animals with a nervous system have structural elements and their morphology that are different from ours. In particular, myelination of axons began relatively late in phylogeny and develops gradually in the ontogeny of each individual. Evolution did not stand still, and engineering solutions developed towards increasing speed, signal-to-noise ratio, and expanding the frequency range as the basis for the development of an adaptive model of reality. It needed spatial, temporal and energy efficiency, as well as resistance to wear and tear.

We do not have wireless channels inside since we are filled not with air but with biological material. We also do not have internal channels that transmit the light frequency range. They would be faster, but the brain uses what it has. In the context of the twisted pair engineering analogy, some of the "oddities" of the configuration and morphology of neural networks become a little clearer. This is a cable network, not power lines. It conveys information, which is a little more complicated than supplying energy through wires. None of the details of axon construction are redundant (not even hundreds of myelin layers). Moreover, if there is a violation of the ratio of the parameters of different structural elements, then this inevitably affects the quality of communication and manifests itself in disorders of the Mind, which we will consider separately.

The bodies of neurons are also a conductive medium for waves of representations. But this is also the coding part of the system, in which messages form: the process of analog-to-digital conversion, coding/decoding. And dendritic "trees" are actively involved in the formation and transmission of messages. But if we are talking about the white matter, then even studies of the brain oscillatory processes are not yet looking at the wave processes in such a large part of the brain. The emphasis is on contacts (synapses) and their "strength," on the speed of the impulse in the "wires" of axons, on the simultaneity or average rate of firing. All these phenomena exist in the system, but they are not the main part of encoding and transferring the code.

Every engineer knows that both the way information is created and how it is effectively communicated are important. But the issue of time for a long time (intentional tautology) did not bother the mainstream of neuroscience at all. When it came to it, researchers began to consider how discrete spikes would fit into existing wires in time to pass without "traffic jams." Of course, it is essential to get into the "elevator" on time, and it must arrive on time. But it is equally vital that it delivers you where you want and at the required speed. If the codes in the system "walked the stairs," then the multicellular living system simply would not survive. Rather, it just would not exist. The solution of related issues of accuracy and speed of information transfer allowed multicellular societies to form and develop.

Usually, the speed of impulse transmission along the axon wires is explained by their good insulation (myelination) and the mechanism of the signal "hopping" along the axon from one break in the myelin sheath to another. The gaps are called nodes of Ranvier, and the process is called saltatory activity (from the Latin saltare — to hop). The importance of such an organization of the signal transmission process cannot be underestimated. Myelinated axons increase the speed and efficiency of transmission by orders of magnitude. For example, an unmyelinated 500 mm squid giant axon requires 5000 times more energy and takes up about 1500 times more space than a 12 mm myelinated nerve in a frog.

Conduction velocity in myelinated fibers is proportional to the diameter, while in unmyelinated fibers, it is proportional to the square root of the diameter. It is a huge advantage in saving space, time and energy while increasing the transmission speed. One can imagine what size the human nervous system would be if there were no myelination. For example, the spinal cord would have to be the size of the trunk of a large tree.

On the one hand, myelin is an electrical insulator that protects against interference and prevents loss. On the other hand, with all the massive number of layers of this insulation, there are gaps. This is a bizarre technological solution when you look at axons as wires. In artificial wiring systems, there is no analogy for such a solution. What's the use of protection if there are holes in it? But if we proceed from the hypothesis about the wave nature of the process, then a lot becomes clear. And then analogies with engineering solutions in wave transmission are immediately found. Nodes of Ranvier can be compared to repeaters allowing communication that can be otherwise obstructed due to distance or interference. The repeater receives waves with particular parameters, modulates them, and re-transmits the information.

Let's look at the nodes of Ranvier from this perspective. First, the internodes (myelin-covered segments of the axon) vary in length but they are substantially longer (millimeters) than the nodes (microns or one-thousandth of a millimeter). It means that the repeaters are positioned at a varying distance as in any wave communication system. They should be not too far apart to keep the information flow but not too close to preserve the cost efficiency. Second, the axonal membrane is open to the extracellular space at the node. This is a technological necessity: the repeater is not a passive relay but a modulator. Third, the structure of the node and the flanking paranodal regions is distinct from the internodes and has a high concentration of ion channels, neurofilaments, vesicles, and other organelles. From the standard wiring paradigm, this seems like a "bottleneck" that only hinders signal transmission. From the wave propagation point of view, this fine structure is tuned to synchronize with the incoming wave and transmit it along the communication channel. As a result, the movement of the wavefront occurs purposefully, swiftly, and with the preservation of the message inherent in it.

Hypothesis:

Axons of neurons, the white matter of the nervous system, is a medium for transmission, reception, integration and disintegration of wave patterns of representations and their synchronization pattern. It is the carrier medium of what we call thoughts, images, sensations, feelings, movements, and the Mind (Consciousness, Psyche, Soul) in general.

From the point of view of physics and technology of the process, this medium and the way it is organized can be called waveguides with repeaters which provide not only efficient and fast transmission of information in the form of wave patterns but also the preservation of their structure, amplitude-frequency and phase characteristics carrying the information-code created by the structures of the neuronal bodies (gray matter). The wave signal "hops" along the axons as an oscillation with specific parameters that carry information transmitted along the waveguide. This process involves a vast number of oscillators at the intracellular level (microtubules, neurofilaments, ion channels, etc.), which are the active carrier for the propagation of the wavefront, the speed of which depends on the energy parameters of the source oscillator (neuron body) and all participants in the "relay race."

Violations in the white matter affect the work of consciousness no less than violations in the work of the gray matter. We will consider pathologies in a special part of the study, but now one illustrative example can be noted: multiple sclerosis. The name reflects the diversity of lesions of the myelin sheath and disturbances in the communication channels' conduction capacity leading to devastating consequences for the functioning of the Mind in all manifestations, from cognitive to motor. Neuroscience drew attention to this critical element of the system relatively recently. But there is more and more research on the active role of this medium and the influence of changes in it on brain function. For example, studies show changes in myelin structure and other changes in the white matter when performing tasks involving the acquisition of new motor skills and recent cognitive decisions.

Myelin formation depends on the activity of glial cells as suppliers of the required substance, but it is determined by direct patterns of activity in the axons.

When, for example, genetic modulation of laboratory mice was used to create individuals in which the function of myelin formation was blocked, they experienced problems with learning and performing tasks. In the normal control group, active cell division of oligodendrocyte progenitor cell (OPC) and myelin formation took place: after 4-6 days of training, up to 40% increase in the number of dividing OPCs (McKenzie et al., 2014). The difference between the experimental and control groups was observed as early as 2 hours after training.

Longitudinal studies of humans show that myelin begins to degrade with age, affecting overall cognitive and motor function (Engvig et al., 2012). Observations of the current process using fMRI showed changes in white matter in humans as early as 2 hours after mastering new situations (Hofstetter et al., 2013). As a result, after a hundred years of concentration on gray cells and their synapses, it turns out that there is a whole "unplowed field" where fascinating and important processes take place.

Let's look at the rest of the network from the perspective of the proposed hypothesis of white matter as an active waveguide. We are left with the dendrites and the body (soma) of the neuron. We have already considered the hypothesis that the body, with all its membrane ion channels, pumps, receptors, and the entire intracellular apparatus, is an operational amplifier, modulator, resistor, capacitor, converter, sampler, quantizer, integrator, and, in general, an active element in the process of signal coding in a hybrid analog-digital chain. It forms messages, transmits and receives them. It sounds like an axiom, but the "devil" is in the details. The most important question is how it works physically.

The central hypothesis of the mainstream is that the entire neural network is a set of electrical wires, and the bodies of neurons simply "integrate and fire" the signal further along the wire. This idea simplifies the process so much that it becomes a parody. It's like a caricature: some details are exaggerated so that nothing else remains. At best, these models are idealization and abstraction that contradicts empirical data if we begin to delve into the process. Anatomically, a neural network looks like a set of wires, and there is no doubt that electrical signals "run" in it. This is part of the truth but superficial. While we "float" on this surface, we run into dead-ends of engineering questions about the essence of the code, the efficiency and speed of its transfer. But as soon as we look at the neural network as an active anisotropic environment for the formation and propagation of waves, many puzzles become clear, including the role of the neuron body and dendrites in the process of receiving/transmitting the information.

For many decades, the process of receiving a signal by one neuron from numerous presynaptic neurons has been described as their "weighing" and summing (this is the whole point of models of neurons-perceptrons according to the integrate and fire principle). The difference between theories is only in the details of how a neuron performs such an operation. In this scheme, dendrites are considered passive wires.

Here is an opinion: "In terms of signal propagation, dendrites behave like electrical cables with medium-quality insulation. As such, passive dendrites linearly filter the input signal as it spreads to the site of initiation, where it is compared with the threshold" (London, Hausser, 2005). Again, there is some truth in this. In terms of conductivity, dendrites are really not the best quality wires. But maybe this is their advantage? The paradox is that if we approach them as passive wires, then their physical parameters clearly do not correspond to the task assigned to them in the framework of such a model. And it turns out that the same question wanders from publication to publication: how does a slow substrate ensure such fast work? And there is no answer to it. Perhaps we are mistaken in the hypothesis and project the wrong model if it repeatedly leads us to a dead-end and diverges from the introjected signals (empirical data)?

Let's try to imagine that the substrate of our brain physically corresponds to the tasks facing it. This is quite a working hypothesis, isn't it? Otherwise, this substrate would not exist, and we would not live. It remains to reconsider our view of the functions performed by this or that element of the system. If the dendrite is only a passive wire, through which an electrical signal must be transmitted quickly and without loss, then it does not meet the requirements. Then we turn the design from head to feet so that it is logical and corresponds to the observed phenomena and not our abstractions and simplifications. Dendrites do not have to be perfectly insulated wires because their role is not to receive an impulse and transmit it passively. They have a very complex topological and morphological structure. If we pay attention to it from the point of view of the participation of dendrites as filters in the process of not only reception but also signal transformation, then a lot falls into place as if by magic.

Dendritic trees of neurons have an active role in the process. But even the fact that these are passive receivers is also true. Simply by definition: any receiver has a passive-accepting role. And even in this passive role, it may not be a simple wire but a part of the computational process, a nonlinear filter. Since dendrites have a network of synapses distributed throughout the space of the neuron's dendritic tree, this spatial organization itself carries a filtering potential. The signal parameters received by the soma of the neuron depend on where the synapse is located on the dendritic tree. Thus, dendrites can mark the signal depending on the delay time and encode it using temporal order. This point was appreciated back in the 1960s, and the idea formed that dendrites carry part of the computational functions. But in general, they are still a poorly understood part of the neural network.

There are two reasons for this: conceptual and technological. But technological problems can be solved if a task is set. For example, many extracellular methods of measurement are not suitable for differentiating the work of dendrites since they mix signals from different elements. But intracellular ones are also problematic. Even thin electrodes and patch-clamp technique do not provide accurate measurements since they rupture the membrane and disrupt the vital activity of the cell, limiting the study time of the process and distorting data in vivo. The use of such methods also requires anesthesia and immobilization of the subject, which means a change in normal neural processes and the inability to study the neural network in conditions of active behavior. They are generally not applicable for thin dendrites.

One experiment set out to overcome these limitations (Moore et al., 2017). The approach turned out to be original since the extracellular method was used to study the intracellular process. The authors assumed that if you install a tetrode (a set of four thin electrodes) at a distance from the cell, then part of the dendrite may end up between the tips of the tetrode. As a result of the work of glial cells, which begin to encapsulate foreign objects, the tetrode will end up inside the dendritic tree system. Gradual implantation will lead to the fact that it will be possible to measure the potential of the dendrite membrane without disturbing it and even in a freely moving animal (rat). The task is set, the job is done. There were many unsuccessful attempts, but the researchers finally measured the dynamics of fluctuations in the potential of dendrites for a long time (up to four days). The results are as follows. Technological: confirming the assumption that this measurement method is applicable. Conceptual: the long-standing idea in neuroscience that spikes of the neuron body are the only element of the computational process was refuted; it was confirmed that dendrites are not passive wires but perform an active computational function. Which one?

The authors found that dendrite action potentials (DAP) frequency is several times higher than the frequency of somatic spikes. It is modulated by subthreshold fluctuations in the dendrite membrane potential (DMP). "DAPs affect synaptic integration and endow neurons with greater computational power and information capacity by turning dendrite branches into computational subunits with branch specific plasticity" (Ibid). Subthreshold DMP oscillations strongly modulated the DAP frequency within a wide range of potential dynamics from almost zero to over 100 Hz. The lag between well-tuned DAP and DMP was close to zero.

The authors asked: "Do DAP and DMP contain information about instantaneous behavior?" They did experiments and found that "pyramidal somata and dendrites both code for egocentric movement, but with differences that illustrate potential computational principles within a neuron" (Ibid). What are these principles? There is no surprise that the authors come to the obvious conclusion about the analog-digital chain. "DAP firing rates are modulated in a graded fashion by the subthreshold DMP. This endows the dendrites with an analog code defined by the depolarization level of the dendrite ... The digital coding, carried out by DAP, coexists with subthreshold modulation, indicating a hybrid, analog-digital code in the dendrites ... The intermediate integration step performed by the dendrites is likely a crucial one in neurons with extensive dendritic trees to allow inputs at distal tufts to be integrated in the somatic response ... The hybrid analog-digital code in a dendrite, generated by a local cluster of excitatory synapses, may be a fundamental building block of neural information processing" (Ibid).

The authors observed the process of signal reception and transduction in a hybrid analog-digital path of a neuron, including an active element at the input (dendritic tree). They do not use the concept of synchronization and concentrate on the computational aspect of the observed phenomena. But, in fact, they describe a wide amplitude-frequency range of the dendritic network, which contributes to the possibility of tuning different neural populations to each other and creating synchronization regions. Note that the low frequency and high amplitude subthreshold DMP (dendritic base pulse) modulates the high frequency and low amplitude DAP waveforms.

This tuning within the dendrite leads to optimal phase coupling (the time lag tends to zero). Regulation of the spatio-temporal parameters of different oscillatory subunits of the entire system of the dendritic tree allows it to carry out its primary function: reception and transformation of various incoming signals into information (frequency-phase pattern), understandable for the central element (soma), which carries out further processing, integration and transmission of data. The details of the ion channels of dendrites are no less complex than the components of the soma membrane, which leads to nonlinear interactions in space and time between multiple incoming flows.

All this creates opportunities for fine-tuning of dendrites as part of the signal reception-transformation-transmission process. And this process is passive-active. In the dendrites, the same process of activation/inhibition occurs as the basis for the regulation of oscillatory activity. Dendrites don't just pick up the signal; they process it. Moreover, the direction of flows in dendritic trees is **not** one-way from the synaptic inputs to the soma of the neuron. The hypothesis that flows go in one direction according to the reception-transmission principle existed for a hundred years and was based on external anatomical observations.

Ramon y Cajal, a pioneer in neural network research and a Nobel laureate, formulated the "law of dynamic polarization," according to which information in the nervous system flows in one direction: from the dendrites to the soma and the axon (Cajal, 1911). Only recently, it became clear that ionic streams in dendrites create independent action potentials that can move in both directions (backpropagation potential). The neuron as a whole and its receiver, in particular, are not a linear signal-reception-transmission chain but a system with feedforward-feedback loops. Again, we face the fact that each element is a nonlinear oscillator and works according to the PAAL algorithm. And it turns out that the pattern of neuron activity depends not only on the processes in the soma but also on the setting of the receiving part of the chain.

Now let's try again to approach this chain technologically and functionally, taking into account the physics of the process. If a neuron's body can be considered a receiver-encoder-transmitter of wave patterns, it must have an active receiving antenna. If there are a receiving antenna and a transmitter, then an active transmitting antenna is also required. Imagine that a neuron is a radio station, a radio receiver, and a relay at the same time. We can remove the word "radio" since we are talking about a different frequency range, but the physical and technological essence is the same. It remains to understand what is the "ether" through which the "radio signals" of the neural network of the brain propagate. This ether is the neural network itself and the entire surrounding substrate.

Hypothesis:

A neuron, as a physical oscillatory system, is a receiver-encoder-transmitter of wave patterns. The dendritic network is both a wave propagation medium and an active receiving antenna, functionally connected to the neuron body in forward-

feedback loops of the PAAL algorithm and participating in the signal decoding process. Axons play the role of a propagation medium (waveguide) and an active transmitting antenna. The body (soma) of a neuron is both a wave propagation medium and an active analog-discrete-analog encoder involved in the creation, transformation, modulation and integration of wave patterns as representations of environmental signals and forming a model of reality.

What does the standard model of a neuron, as a producer of discrete identical "shots" and an adder of incoming signals, offer as a role for dendrites? The authors of the review cited earlier note that there are four main versions: synaptic scaling, synaptic amplification, local spikes, and global dendritic spikes (London, Hausser, 2005).

In the first model, conduction delays from spatially distributed dendritic synapses are scaled with distance, and thus everyone gets an equal "voice" in such a "dendritic democracy." There is no direct experimental confirmation of such an equalization. The second hypothesis of subthreshold amplification implies that internal voltage-dependent dendritic fluxes can amplify synaptic inputs en route to the soma, compensating for their attenuation due to distance. The third model assumes the presence of local dendritic spikes that occur during the interaction of different input signals and regenerate the strength of the flow. The model of global dendritic spikes is, in fact, about the same but speaks about those neurons where the length of the dendrites is so great that the absence of a compensatory mechanism would lead to complete attenuation on the way to the soma. Experiments show that there are secondary spike initiation zones.

So, the first model says that there is no variability in the internal parameters of the neuron body spike, and it produces identical "ones." The analogy with a digital code does not work since no "zeros" are considered. In such a model, the pattern consists of "shots." Since the only way to make a pattern from ones without variations and other code elements is the variability of the speed of these identical "shots," firing rate change is considered to be a neural code. As we have already noted, it does not correspond to the realities of the system operation or technological common sense. One erroneous hypothesis pulls a chain of other erroneous ones.

If the body of the neuron fires identical shots, then all the signals entering it must be summed up so that there is no variation in the incoming stream. The result is an "integrate-fire" model. Everything should be summed up linearly. But since there are differences in spikes, it is necessary to apply a "one size fits all" approach so that the incoming signals from different presynaptic neurons throughout the dendritic tree would have the same voice. It turns out that the entire complex structure of the dendritic network was created to bring everything into identical shots with a single variable parameter of speed. The hypothesis of leveling the difference in fluxes has no empirical evidence. Still, it tries to put a basis under the already existing hypothesis of identical spikes and tempo code, i.e., a hypothesis serves as a "proof" of another hypothesis. The rest of the hypotheses about the role of dendrites state the presence of subtle mechanisms of signal modulation, but everything is reduced to amplification (more precisely, overcoming the attenuation). They talk about the amplitude and signal retransmission regulation, which should exist in any version of the code. And it is no coincidence that there is experimental evidence, but this is only a small part of the process.

Let's look at a schematic representation of different types of "dendritic trees":

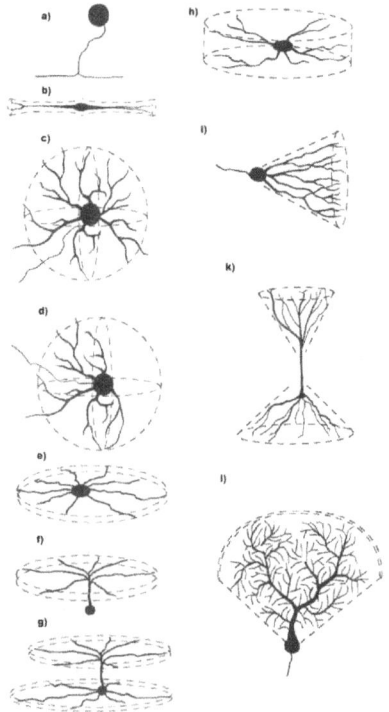

Visually, we can compare this with trees, corals, or something else from biological reality. But let's look at them from the point of view of technological reality and physics of the process in our system. Then the name "dendritic antenna" will be suitable. And not because the shape resembles different types of antennae, but because the processes in the dendrites speak of a particular function, which in turn affects the shape. The proposed term does not come from external analogy, as the previous ones, but from the internal technology of the process. Antenna's shape has technological meaning.

Such an analogy corresponds to the physics and physiology of the process and in no way contradicts the previous analogies but clarifies them. The comparison to trees and corals, that system anatomy researchers have thought of, reflects external form. But the form is usually associated with function. Trees and corals are also, in this sense, antennas that pick-up waves of external signals and their shape creates an effective organization of such a process. The first artificial antennas were designed by Heinrich Hertz in 1888 to prove the existence of electromagnetic waves. He was a member of the "Wave" team, and it is no coincidence that the unit for measuring energy fluctuations is named after him. We continue the game of this team, and the name "dendritic antenna" goes deeper into the physics of the process in the neural network.

Neurons are not the same, and their antennas are also very different. For example, the antennas of the pyramidal cells of the cortex are very long:

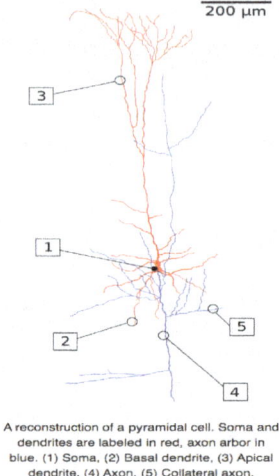

A reconstruction of a pyramidal cell. Soma and dendrites are labeled in red, axon arbor in blue. (1) Soma, (2) Basal dendrite, (3) Apical dendrite, (4) Axon, (5) Collateral axon.

Antennas of Purkinje cells in the cerebellum are branchy. Drawing made by Ramon y Cajal:

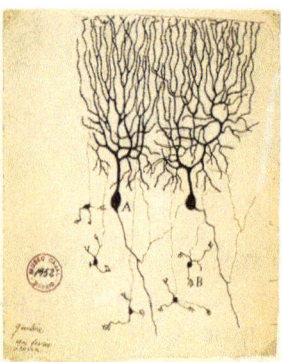

The antennas of the retinal neurons are relatively short and sparsely branched:

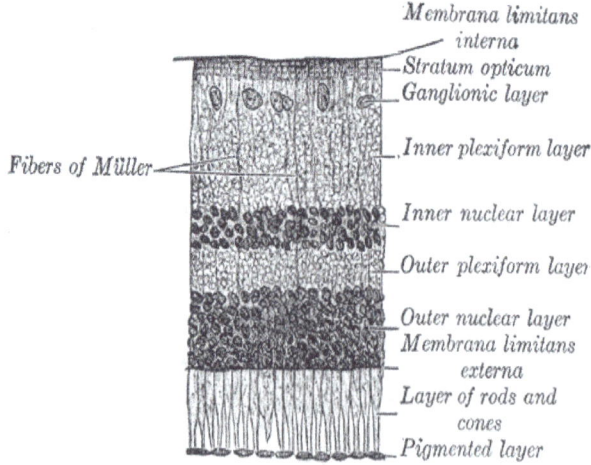

Why such a difference? The answer is most likely in the functional difference of the neurons themselves: some are participants in higher integration and capture a vast number of patterns and a wide frequency range; others are very specialized converters tuned to a narrow band. Here we will have to restrict ourselves to a very general hypothesis that different forms of dendritic antennas reflect various functional tasks of neurons, and the antennas themselves may differ in type. Perhaps they can even be categorized, like antennas in artificial communications: traveling-wave antennas, resonant antennas, broadband, narrowband, low-frequency, high-frequency, and so on. The dendritic antennas of the neurons of higher integrators may be traveling-wave antennas with a wide reception band, and the transducer antennas are resonant antennas with a narrow band.

Since until recently, dendrites were approached as passive wires, at this stage, these networks are so little studied that it is hard to find experimental data in sufficient detail and look at them from a new point of view. As the authors of the review mentioned above note: "Ultimately, understanding the role of dendrites in neural computation requires a theory. This theory must identify the benefits of having dendrites and reveal the basic principles used to provide these benefits" (London, Hausser, 2005). A new look can generate research and fresh data.

For a long time, dendrites have been considered a channel for one-way flow from presynaptic neurons to the soma. But then it turned out that there are not only axo-somatic and axo-dendritic connections but also dendro-dendritic synapses between dendritic trees of different neurons. It was even called a "quiet revolution": this morphology was in direct conflict with the paradigm of the direction of flows in the network and the paradigm of the dendrites function.

Moreover, it turned out that such dendro-dendritic synapses are reciprocal, i.e., both neurons in such a synapse become presynaptic depending on the situation. A logical conclusion was made: the dendrite directly transmits information to other neurons. But what is this information? Based on the hypothesis proposed here, we can assume that the antennas of the neuronal population are connected into a network for wave pattern receivers for processing, creating and transmitting representations.

The location of synapses on the dendritic surface is not chaotic. They have places of accumulation and distribution according to the activation/inhibition function. Usually, inhibitory synapses are located on the trunk of the tree (the antenna axis), which allows several inhibitory pulses to modulate the parameters of the entire stream. The modulation mechanism works, the main driver of which, however paradoxical it may sound, is inhibition. The presence of feedback from the soma and axons allows modulation of the dendritic active antenna's parameters. The entire system works as a single receiver-encoder-transmitter within its "powers": frequency bands and functions in signal processing. The dendritic antenna has all the necessary tools: ion channels and pumps work there as regulators of energy fluctuations. The patterns of neuron activity depend both on the processes in the soma and the dendritic antennas.

Regardless of whether it is receiving or transmitting, an antenna performs the function of interacting with energy external to it and converting incoming oscillations into outgoing ones. Passive antennas simply receive vibrations, and the parameters of this reception depend on their shape and material. We can say with a high degree of probability that there are no purely passive antennas in the nervous system.

An active antenna implies the presence of additional devices: amplifiers, attenuators, decoders, and so on. This, of course, requires constant expenditures of energy and technological complications. But they justify themselves since the spectrum of signals, the range of adjustments, compensatory and other possibilities increase significantly. If we accept the working hypothesis that dendrites are active receiving antennas, we have to investigate wave reception and regulation processes.

Despite the lack of attention to dendrites over the years, a body of data has accumulated in neuroscience about how this network works. No one (as far as the author of these lines knows) has yet called dendrites antennas, but they are often mentioned together with the terms "oscillations" and "traveling waves."

Take, for example, an article that is directly related to the topic of this chapter (motion technology). Its name speaks for itself: "A dendritic mechanism for decoding traveling waves: principles and applications to motor cortex" (Heitmann, Boonstra, Breakspear, 2013). The authors begin with these statements: "Traveling waves of neuronal oscillations have been observed in many cortical regions, including the motor and sensory cortex. Such waves are often modulated in a task-dependent fashion although their precise functional role remains a matter of debate. Here we conjecture that the cortex can utilize the direction and wavelength of traveling waves to encode information. We present a novel neural mechanism by which such information may be decoded by the spatial arrangement of receptors within the dendritic receptor field. In particular, we show how the density distributions of excitatory and inhibitory receptors can combine to act as a spatial filter of wave patterns. The proposed dendritic mechanism ensures that the neuron selectively responds to specific wave patterns, thus constituting a neural basis of pattern decoding. We validate this proposal in the descending motor system, where we model the large receptor fields of the pyramidal tract neurons — the principle outputs of the motor cortex — decoding motor commands encoded in the direction of traveling wave patterns in motor cortex. We use an existing model of field oscillations in motor cortex to investigate how the topology of the pyramidal cell receptor field acts to tune the cells responses to specific oscillatory wave patterns, even when those patterns are highly degraded" (Ibid).

A very promising statement, and if you read the text from a technological point of view, then it remains to add: this is an attempt to describe dendritic trees as antennas and the whole chain as reception-decoding-encoding-transmission.

What did the authors do? They created a computer model that simulates the responses of the neurons in the motor cortex to incoming waves. The dendritic tree's receptor field is modeled as a spatial filter similar to the Gabor filter, which selectively triggers the action potentials of the neuron upon detection of a particular wave pattern. The authors refer to previous attempts to model cells of the visual cortex as Gabor filters and assume the possibility of applying the same model to the motor cortex. It should be noted here that the Gabor filter in artificial technologies is a linear filter for analyzing the frequency components of images, and it was naturally used as an analogy for modeling the operation of the visual system.

Mathematically, it is a convolution of the Fourier transform of sinusoidal components and the transformation of the Gaussian normal distribution function. In signal processing, it is helpful in that it allows to isolate components where there are sharp differences: text analysis, face analysis, fingerprint analysis. It can analyze the signal where a clear pattern has to be identified and converts the spatial phase shift to temporal shift. In general, it is a very apt analogy for the operation of a dendritic antenna.

The authors suggest that "dendritic fields in cortex may serve as biological Gabor filters of internally generated patterns of oscillatory activity," "exploit asymmetries in the spatial profile of the dendritic receptor densities in order to advance or retard spike timing relative to the incoming oscillations," "neurons of motor cortex may use Gabor filtering to translate those oscillatory patterns into steady motor output in the spine." (Ibid). The principle is the same for the receiving and the transmitting antenna. Why not? Enough data have accumulated that dendrites are not simple input adders, as McCulloch and Pitts presented them in their linear neural network model (McCulloch, Pitts, 1943). They work as active structures for the allocation of spatio-temporal patterns. In such work, determining the frequency and phase parameters of the incoming oscillations is very important: it is the function of the antenna that picks up the waves.

The authors took as a basis the waves in the beta range (approximately 12-30 Hz), which have long been associated with planning and implementation of movement but only recently began to be considered as waves with a temporal-spatial structure. They modeled the cortex's neurons as Kuramoto coupled oscillators. Kuramoto model has certain assumptions: the identity of the oscillators and the sinusoidal dependence of the interaction on the phase difference of each pair of oscillators. Such a network model assumes only phase coupling, which significantly reduces the analysis, but you have to start somewhere.

In the simulation, it turned out that a population of pyramidal tract (PTN) neurons with identical reception parameters samples the waves, and a population of motor neurons converts outgoing PTN signals into muscle movement. The model showed that this movement could be controlled by varying the cortical wave pattern so that the orientation of the pattern and the receptive field of the PTN are interrelated. The topology of inhibitory connections regulates the length and direction of the emerging traveling wave.

The authors modeled synaptic fluxes while simulating the "bombardment" of the receptor fields of dendrites by propagating waves of cortical activity. There is a caveat: the spike patterns were simulated as random shooting with the Poisson distribution of random and independent variables. We will consider the disadvantages of this approach to the rhythms of the brain later (see "Part Six. Harmonies of the Mind"). What is important for us now is the idea that receiving and decoding an incoming signal is an analysis of wave patterns. In the experiment, the phase of dendritic oscillations was determined by the propagation of the wave pattern along the receptor field, i.e., the dendritic antenna was synchronized with the incoming oscillations and thus could read the pattern.

"It was found that individual PTNs responded selectively to cortical wave orientation but the response rates were limited to discrete frequency bands (10, 20, 30, 40 Hz) due to entrainment by the intrinsic 20 Hz oscillations in cortex ... The sinusoidal input forces an oscillation in the somatic membrane potential which is phase locked to the input. The resulting spikes tend to coincide with the rising peaks of the injection current although the number of cycles required to trigger a spike varies with current amplitude. Injection currents below 0.50 nA fail to elicit any spikes (not shown) whereas currents of 0.50 nA and 0.51 nA elicits spikes every third and second cycle respectively. At 1.00 nA the oscillating current produces regular spikes on every cycle and at higher currents (e.g. 1.50 nA) double-spikes appear. The somatic responses to the full range of possible current amplitudes resembles the 'devil's staircase' of a periodically forced resonator with major frequency plateaus at 20 Hz and 40 Hz interspersed with minor plateaus at 10 Hz and 30 Hz and a multitude of even smaller m:n phase locked solutions between those. The corresponding response frequencies of the somatic compartment were predicted by mapping the dendritic tuning curve onto the somatic response to pure 20 Hz input. The result is the likelihood of the PTN firing at each of the four dominant entrainment frequencies (10 Hz, 20 Hz, 30 Hz, 40 Hz) for any given wave orientation in the cortex ... It is found that the distributions of frequency-specific responses are balanced so that the combined spike output of all PTNs is itself a smooth function of wave orientation" (Heitmann, Boonstra, Breakspear, 2013).

If this is not about synchronization in the process of receiving incoming waves, then what else is it about? It is worth emphasizing here once again that the authors only made a biologically plausible simulation with some assumptions (strict periodicity of a sinusoidal signal, the identity of oscillators) and an arbitrary choice of oscillatory parameters (frequency, amplitude, phase) that are within a known range but may differ from real parameters in the live network. The authors are forced to ignore the variability of the input signals with rhythmic detail, which is very important for the process. They explain this approach: "Neurons exhibit variable inter-spike intervals in vivo that are difficult to replicate in purely deterministic models" (Ibid). Since neuroscience has not even approached the analysis of actual brain rhythms (sequences and durations of "sounds and pauses" of the neural code), this approach is justified in the intermediate model. It shows that the reception and transmission of a signal inside the neuron are based on the synchronization of the dendritic antenna with the incoming waves. It can be called an illustration of how this can happen.

However, for a realistic description, we cannot do without a holistic theory and a detailed study of the processes from the point of view of the requirements of such an approach. A physically based theory will indeed require the same physical and physiological basis from research for its confirmation. This is the lemniscate of cognition: a model requires empiricism for testing, and empiricism requires a coherent model for its "placement on the shelves." As the authors regretfully note, "there is relatively little research exploring the potential of dendrites to discriminate spatial patterns of oscillatory inputs" and "the spatial organization of those oscillations as traveling waves is only a recent discovery" (Ibid). The authors' article convincingly shows that the dendritic receptor field is a filter of spatial patterns in the phase of incoming oscillatory signals. "Dendritic computation is thus portrayed as the integration of synaptic phases rather than the integration of synaptic membrane potentials ... This phase-based approach is consistent with emerging evidence that dendritic integration is sensitive to the relative timing and spatial location of synaptic input on the dendritic arbors" (Ibid).

It remains to apply the word "antenna," and the technological picture develops into a coherent hypothesis that smoothly fits into the general concept. This makes it possible to explain the phenomenon, which the authors took outside the model framework: "The present model does not account for decoding the large-scale wavelengths (1 cm) observed in motor cortex since such wavelengths far exceed the spatial resolution of individual PTN receptor fields" (Ibid). If we consider dendrites as antennas connected in a common chain (remember the presence of dendro-dendritic two-way synapses), then there is no technological problem in that PTN populations can receive, decode, encode and transmit large-scale wave patterns. The authors themselves noted that "pattern formation and discrimination remains feasible even when the intrinsic oscillation frequencies are broadly distributed" (Ibid).

In their model, the authors considered only one parameter of the propagating wave: orientation. They admit that it is a deliberate simplification. Indeed, waves travel in many directions and carry other parameters that are likely to be of importance to the receiving antenna. The authors' task was to show the possible mechanism of this technology at least in the context of one parameter, and they "anticipate that dendritic trees are capable of filtering a broader class of oscillatory spatiotemporal patterns." They also note that "traveling waves are not restricted to motor cortex and the proposed dendritic mechanism may also generalize to traveling waves in other modalities" (Ibid).

The approach to neural activity as waves is not new. One of the pioneers of neurophysiology, Charles Sherrington, wrote: "Imagine activity in [the brain] is shown by little points of light. Of these some stationary, flash rhythmically, faster or slower. Others are traveling points, streaming in serial trains at various speeds. The rhythmic stationary lights lie at the nodes. The nodes are both goals whither converge, and junctions whence diverge, the lines of traveling lights" (Wu, 2008).

It was a dream: a dream to see waves of the Mind. In a sense, we are getting closer to this dream. Technology advances, and sometimes visionaries' dreams come true almost literally. Now, thanks to voltage-sensitive dyes (VSD) techniques and optical technology, we can look at the stationary and moving lights in the brain. They allow tracking changes in membrane potential, speed and direction of its movement. They even penetrate the cell without violating its integrity. They make it possible to measure spatial and temporal variations in the membrane potential on the surface of one cell and population. It is, of course, not an ideal, and there are, as usual, disadvantages: there is no possibility of a purposeful study of the desired area of the membrane (the problem of staining); low level of signal-to-noise ratio, requiring additional algorithmic filtering, which can obscure actual processes; despite being mildly invasive, it disrupts cell function and can lead to long-term adverse pharmacological effects.

These are all technical difficulties that can be overcome over time. But there are also conceptual complications. Oddly enough, the mainstream continues to look at waves but see discrete spikes. Although there is nothing strange about this: we often see what we want to see (projection in action). Thus, the authors of the review titled "Traveling waves of activity in neocortex: what they are, what they do" wrote in the introduction: "These waves provide subthreshold depolarization to individual neurons and increase their spiking probability. Therefore, the propagation of the waves sets up a spatiotemporal framework for increased excitability in neuronal populations, which can help to determine when and where the neurons are likely to fire" (Wu, 2008).

It turns out that waves are a way to increase the probability of a discrete "shot," which contains all the information. Of course, the wave must relate to the phases of oscillations of the elements participating in it. By definition, it is the propagation of changes in the physical parameters of these oscillations, including the phase. But the fundamental question is: is the meaning of the wave in the creation of the peak phase of the oscillation (spike) or the propagation of energy-information as a pattern for changing such physical parameters? Rather the second than the first. We again face a "caricature": a very bright discrete part of the process stands out, and it is argued that this is the meaning of the process. And the cause and effect are confused: it turns out that the wave creates oscillation phases, but not oscillations and their phase dynamics create a wave.

Interestingly, the authors compare the observed brain waves to waves in a stadium where participants stand, raise their arms, and sit down. They even stress that "during a stadium wave, each individual is only required to rise slightly after the person immediately next to him/her and does not have to fully stand up to participate in this mass phenomenon. Similarly in the cortex, mild depolarization during the wave increases the chance of spiking in the population and these spikes will in turn depolarize more postsynaptic neurons in the neighboring area to sustain the wave propagation. Propagating waves may contribute to cortical function in a number of ways. First, waves provide a background depolarization to a selected cortical region ... Second, a sensory-evoked wave propagating to a larger area would increase the sensitivity/network gain for incoming stimulation ... Third, propagating waves associated with an oscillation can organize spatial phase distributions in a population of neurons" (Ibid).

What are all these "contributions" of waves to brain functions for? If we take the authors' analogy with a wave running through the stadium, then we get a picture turned upside down again. The authors' assertions that waves only increase the likelihood of spiking, create background depolarization and increase the sensitivity of neurons are similar to saying that the meaning of a traveling wave in a stadium is to create an opportunity for each participant to jump up. But everything is precisely the opposite: the participants in the process rise and sit in order to create a wave in which the meaning is embedded. In this example, the meaning is simple, like the wave itself: we all enjoy what is happening here.

Brain wave patterns are much more complex and varied than a wave at a stadium. But the essence is simple: the activity of each specific neuron, whatever the form of its phase portrait (whether it jumped higher or lower, fully or partly), is tuned to participate in a wave pattern, as a carrier of meaning, and not background activity. It is amazing how a persistent old paradigm makes authors contradict themselves and their analogies and not notice it. And at the end, they add the third option about the phase distribution, but again with inverted logic. It is not the waves that organize the phase distribution, but the phase distribution of the interacting oscillators creates a phenomenon that we call waves.

And for some reason, the distribution of phases for the authors has only a spatial aspect, while it is about time. The authors even contradict the observed phenomena, which they write about: "Spikes from individual neurons cannot be seen, because they are sparse in the population and have very short duration (~1 msec), so that their contribution is much less than that of the long and overlapping subthreshold potentials" (Ibid). But what about the postulated main function of waves, as the creators of a greater likelihood of neuron spiking and determining where and how it will fire?

Indeed, if you look at a running wave at a stadium from a long distance, you will not see the "spikes" of individual fans. You will see the pattern created not only by the peak phase but also by the entire oscillation of each individual. This pattern is obvious and carries its meaning. Moreover, there is no clear threshold to say: this was a "subthreshold oscillation,"" and this is a "spike."" When did the bounce-spike occur? When a person was rising hands or at the full stand-up? The authors themselves noted that there is no requirement to stand up fully to participate in the wave. The same applies to neuron oscillations: this is a continuous process in which discrete phases can be distinguished only conditionally, and the very concept of a discrete spike is perhaps the most enduring myth of neuroscience.

But the presence of a common meaning in the wave does not in any way deny the meaning of the message of each participant in the wave, whatever it may be, even if he did not stand up at all. In the music of the Mind, notes have the same meaning as pauses, and each note and pause has its own parameters and place in the general structure. But for this, they must be coupled-synchronized. The authors quite correctly point out: "In a system with coupled oscillators, waves can travel back and forth like water waves in a pond. Coupled oscillators in the cortex can generate complex spatiotemporal patterns such as plane, spiral, and irregular waves" (Ibid). And this is the whole point of the process of creating meanings.

The phenomena observed using VSD technology are entirely consistent with Sherrington's prophetic vision of nodes and junctions. "Two types of spatiotemporal patterns are observed in oscillatory waves in cortex. One is referred to as "one-cycle-one-wave," that is, there is a clear correlation between the oscillation at each location and waves distributed in space; each oscillation cycle measured at a given location is associated to a propagating wave swapping through that location. The other type does not have a clear correlation between oscillations and waves; usually many oscillation cycles are enveloped into one propagating wave" (Ibid).

But the prophecy works if it is based on a real physical mechanism. Any prophet proceeds from knowledge of patterns: having experience with past patterns allows you to predict future patterns. Indeed, imagine waves in a pond. If you throw one stone (a source of perturbation and oscillation), then there will be a clear correlation between oscillations at a "node" and traveling waves. If there are many stones, then this will be a general interference pattern of the interaction of many oscillations and waves, and the nodes will be junctions. No mysticism, physics of the process. If there is one source, then the frequency, length and phase of the wave will be directly related to it. If there are many, polyphony and polyrhythm begin. One note is not silence, but not music yet; many notes are not music until they create an overall synchronization pattern.

Such patterns are observed "in olfactory, visual, somatosensory, auditory, and motor cortices, (where) VSD imaging or electrode arrays have revealed evoked propagating waves manifested during sensory/motor processes and during spontaneous events. The propagating velocity of these waves is fast, about 0.1 to 0.25 m/sec and the duration of the waves is 50 to 400 msec ... Spontaneous waves in cortex reflect organized, large-scale intrinsic cortical population activity, which may play an important role in cortical processing" (Ibid).

It remains to draw conclusions from observations and decide whether it is just "ripples on the surface of the pond" and background activity or an important mechanism in signal processing and, perhaps, even the essence of the whole process. What answer do the authors give to the question in the title of the article and to which it is devoted? We are no longer surprised in finding the same old line about poorly understood things: "Although almost every sensory cortical process examined by VSD imaging or electrode arrays has revealed propagating waves, the connection between the spatiotemporal pattern and the function of the cortex is still missing" (Ibid).

In short, there is no answer to the question of what waves do. Maybe they don't do anything at all? Within the TTT, there is a straightforward answer: they create the Mind; they are what we call feelings, thoughts, desires, plans, and so on. They also create the embodiment of our feelings, thoughts, desires and plans — movement. The question is how they do it. Let's go on with answering it.

CHAPTER 7

MOVEMENT CONTROL TECHNOLOGY

The brain's control of organized movement gave birth to the generation and nature of the mind.

Rodolfo Llinas

The Mind of a living system works for moving the body and manipulating objects of the environment. But there is one long-standing dream of humans, which was not always explicitly expressed but was implicitly contained in the motivations for the development of all technologies: the manipulation of external objects without the movement of one's own body. As the saying goes: laziness is the engine of progress.

This dream is getting close to becoming true as machines are already doing a lot of work controlled by computers. But the next step is to control computers with our thoughts. Internal Mind movements will directly manipulate objects. In fairy tales, a hero had to pronounce a magic spell. In modern technologies, everything points in the direction that he will not even need to say anything but only think. It is the essence of the task of all modern BCI (brain-computer interface) and BMI (brain-machine interface) technologies.

But the fairy tales accurately reflected the components necessary for this: the will of the subject and a code relating this will with external objects. The nervous system has necessary "magic spells" for the manipulation of body effectors. The neural code creates a connection between intention, goal-setting and fulfillment. Living and artificial intelligence interface is an increasingly tangible reality, but we need a magic spell. In all fairy tales, it was a big secret revealed to the hero for special merits. It is no coincidence that the stumbling block in all research is the brain code.

At the current level of understanding of the system, we can link the activity of neurons and the action of external effectors. To do this, we can use EEG, fMRI, multi-electrode arrays, and so on. They all have varying degrees of practicality and effectiveness. And there is no doubt that technologies for taking data on neural activity will be improved. They can be used without even knowing the brain code. But can we go around the central problem of finding the "magic spell"?

There are two options for approaching the interface problem. The first one can be called biofeedback: the use of the subject's learning of volitional control and manipulation of the parameters of physiological functions to control external effectors that learn to encode these changes in parameters. Theoretically and practically, it can be any parameters: we can record electromagnetic waves in EEG and MEG, BOLD (blood-oxygen-level-dependent) signal flows in fMRI or the average activity of a population of neurons in the area of electrodes. But we can also read changes in the conductivity of the skin, pulse, muscle tone, and so on. These parameters, which with sufficient training can be controlled by purposeful effort and reflect brain activity, are, in fact, the same indirect signals as the previous ones. Even reading of the average activity of neurons with implanted electrodes is not a direct reading of the "spell," but only approaching and touching the "magic spell book."

This method has advantages in both resolution and speed. It is more direct in the sense of its connection with the control system. But the key to this approach is the need to train the subject to purposefully change the parameters of his physiological states and train the effector to perceive these changes as commands for specific actions. If the experiment is carried out with an animal, then very lengthy training procedures are required since it is impossible to explain in words what the experimenters want. But you can train a monkey to change its brain activity to feed itself with a robotic arm (Velliste et al., 2008). You can do the same with a human and much faster (Hochberg et al., 2012, Fukuma et al., 2015, Kim et al., 2015).

The authors of one experiment noted that the main result of their study was that "monkeys could learn to use virtually any motor cortex cell to control muscle stimulation, regardless of the cell's original relation to wrist movement" (Moritz et al., 2008). In other words, it was not that the neural code was connected to the code of an artificial system, but that a living system controlled a particular biological parameter in order to control the artificial one. With the same success, you can associate control with a change in any of the parameters that the system can change by deliberate effort. The scheme looks like a direct connection between the brain and artificial intelligence. Still, in reality, we are talking about indirect control through an algorithm for correlating the parameters of the internal activity of a living system and external actions of an artificial one.

Biofeedback interface connects the subject and external effectors' activity through a learning algorithm of the interaction of the biological and technical parameters. It is almost a fairy tale. But we are still very far from the degree of accuracy, speed and efficiency of the nervous system manipulation of its internal effectors. It is still easier for the hero to do things himself. Unless, of course, he is disabled. For paralyzed people with disabilities, even the ability to control a robotic arm to bring a drink is already a huge step.

You can watch the video and notice the emotions of the woman who participated in the experiment ("Paralysed woman moves robot with her mind" www.youtube.com/watch?v=ogBX18maUiM). However, there are reasons to say that this interface method has minimal capacities compared to a natural brain-body interface. And the point is precisely in the "magic spell," in the requirement to read the brain code, not indirect data about its activity.

The second method of the artificial interface can be called biomimetic approach: the application in technical devices of principles, properties and functions similar to living systems, the combination of biology and technology (bionics). In the previous method, a purposeful effort of the subject is required to change the parameters of the activity of organs. There has to be a learning period since the brain begins to create the necessary representations to control the external effector.

And here arises the question of the efficiency and balance of computational, energy costs and results. When we make ordinary movements, the brain does not use all the higher integrators since this is not justified from the energy use perspective. Moreover, many living systems do not have all the levels of integrators that humans and monkeys (participants in the above experiments) have. But they also move very efficiently.

The bottom line is that the efficiency and speed of the nervous system are so high that real biomimetics (artificial technologies similar to biological ones) require a fundamentally different solution. You need a connection of fully compatible devices, which means that they must speak the same language. Artificial technology should perceive the code and direct representations created by the nervous system when controlling the body and should be able to give feedback in the same language.

The first approach, based on the creation of a new instrument-effector, its training and feedback from indirect parameters, and the second, which assumes more direct methods of communication and the design of a biological function analogue, do not contradict each other. After all, a living system also learns in the real world based on feedback. Not all actions immediately follow an economical chain without the participation of energy-consuming higher integrators. We all know this from the experience of learning any complex movement: at first, you need to strain your brain. As in the experiments with the robot arm and other BCI technologies, the process flows smoothly after the intense training period.

In some ways, this technology is also bionics. The robot's hand is made similar to ours (the mechanics technologies are fundamentally the same). The computer that controls it and is connected to devices that read the brain activity is the same analog-digital device as the nervous system. The point is not how we call this technology and approach. For a fairy tale to come true, these external devices must really read the brain and know its code.

The neural code is still the mysterious "magic spell." But maybe it is enough to develop technologies for mutual learning of a computer and a human based on indirect signals? Is it bad that a paralyzed person can control external effectors, at least at this level? It is excellent. But this is not directly related to thoughts or other representations. The question of how the brain works, how it creates these representations, and how they can be decoded remains open. And here, the fundamental thing is the paradigm from which both BCI theorists and engineers proceed.

The main focus is on increasing the resolution of non-invasive technologies for measuring brain activity or improving the compatibility of invasive technologies with living tissues. Algorithms for processing this activity develop and the range of commands that effectors can execute is expanding. But even if we are talking only about creating controlled prostheses for disabled people, with such an approach, there will inevitably arise, if not a dead-end, then a serious barrier. In such an interface, the artificial effector is controlled according to the principle of pulling a puppet by the strings. This technology cannot reproduce the entire palette of movements and sensory and proprioceptive feedback from the body that influences these movements.

The scheme as a whole is very similar to PAAL: there is brain activity, an effector, a connection between them, movement as a result of feedforward and feedback commands, the process of learning and correction. But the interface remains indirect, not very different from the interface that any other artificial technology has: it is not controlled by thought in the true sense of the word. It is not governed by the representations created by the Mind.

Even voice control is, in this sense, a more logical technology (moreover, non-invasive). After all, what is voice control? It is the decoding by the machine of the code (speech) known to its creator and programmer. All that remains is improving signal conversion technologies and expanding the range of interaction between a person and a machine controlled by a voice. We are almost in a fairy tale. We pronounce the spell, and the door opens, the light turns on, a car goes by itself, etc. The disadvantage is the low speed of information transfer and no connection between the brain code and the effector code.

But the same disadvantages are inherent in the standard BMI scheme. Although the code is unknown, the illusion of a direct connection is created. Why, then, all these tricks with connecting electrodes to the brain, if this is just an illusion of thought control? But nothing happens at once by a wish. Technologies evolve by trial and error and very often by overcoming barriers and breaking out of impasses. Changing the theoretical paradigm is a practical challenge. Just like the correction of the model of reality is a practical task of the brain.

In the review article "Interfacing with the Computational Brain," neuroscientist Andrew Jackson and biophysicist Eberhardt Fetz noted: "Fundamental issue that has received surprisingly little attention within the BMI community is the validity of the primary assumption underlying the biomimetic approach, namely that movement parameters are meaningfully and consistently encoded by the firing rates … On the one hand, the lack of a specific 'encoding' scheme does not render 'decoding' impossible … On the other hand, without principled assumptions about what parameters are encoded, there is no reason to expect any decoder to generalize beyond that sub-space of movements sampled within the training set … These decoders may be inappropriate for fast, on-line corrections of errors arising from central sources of variability in neural command signals ... Human reaching movements are typically fast, accurate and characterized by stereotyped features such linear trajectories and bell-shaped speed profiles. It is commonly accepted that such movements cannot be generated by simple feedback control, due to delays in sensory information and the need to co-ordinate muscles acting across different joints of the limb. Instead the motor system requires advanced knowledge of the kinematics and dynamics of the limb and environment; this knowledge is often described as an internal model ... However, while optimal feedback control explains several features of natural movements, distinguishing the relative contributions of inverse models, forward models and sensory feedback remains an unresolved issue" (Jackson, Fetz, 2012).

The movements of the "puppet" (the standard interface system) are limited, too slow, and inadequate to the realities of the living environment. In the fairy tale "The Adventures of Pinocchio," the doll turns into a living boy. What is required for such a miraculous transformation? Alive Pinocchio possesses consciousness: an internal model with representations of movements. The scheme is simple: projection of the model, introjection of sensory signals and correction. The relative role of each stage is not an "unresolved issue" since it is normally a continuous, synchronized and coherent two-way process. Without a model, there is nothing, but the model is meaningless without testing and sensory information. Lemniscate is a closed and smooth trajectory. Any rupture or bias in this technological chain stages' relative contribution is a pathology of the system.

In this one-actor theater, there is a script and a director (reality model). But is there a "puppeteer," a central source in the nervous system that pulls strings? This metaphor of a puppet has explicitly or implicitly owned the minds of physiologists for more than one century and is still mainstream in neuroscience. In a nutshell, the picture of how the nervous system controls the body in such a paradigm looks like this: there are communication strings (neural circuits) through which commands are transmitted from top to bottom, and reports are received from bottom to top. It is truth, but only part of it. The fundamental question is how the commands and responses are formed and transmitted. It would seem that everything is simple: there is a spike and its propagation, an electrical impulse and its transmission through wires.

However, our "wires" are physically not fast enough to transmit information throughout the body in an almost instant online mode if we assume that control goes as a linear transmission of impulses from top to bottom and from bottom to top. But nervous system does it somehow. This is the paradox of the temporal aspect.

But there is also a spatial paradox in the architecture of this wiring. These "strings" have one property that is striking from the point of view of the linear paradigm. If we proceed from such a management model, then nature made the wiring in our house, which we call the body, strangely: the bundles of wires do not go in parallel paths back and forth but converge and diverge. They converge towards the center, diverge there again, converge again on the way to the periphery, and again diverge at the periphery. At one stage, they are parallel and differentiated, and at another stage, they are sequential and mixed. It is a hybrid architecture. This applies to both motor and sensory pathways.

For example, the cat's olfactory system's convergence index at the first stage of signal conversion is approximately 140 million receptors to 80 thousand mitral cells (second-level neurons) in the olfactory bulb, i.e., it is 1750:1. A simple question: how can these flows be differentiated and integrated at the same time? In neuroscience, this issue is called a binding problem (BP), which has two sides: segregation and combination problems (BP1 and BP2). If we do not know the technology of the brain, we are bound to make assumptions based on the knowledge of how the same problem is solved in artificial systems created by our brain. The task is similar, and physics is common. The difference is only in physiology, i.e., what substrate performs the function relying on the physical processes shared by both kinds of systems.

For example, we are not surprised that our TV receives different programs while connected to one cable. Moreover, all these programs "live" in it at the same time. You can even display many channels at once on one screen. Or we can watch one and record another. This diverse information comes through one cable. How can it happen? Firstly, the signals are encoded and separate representations are created, and secondly, these representations are distributed over different channels-frequencies. All that remains for the receiver is tuning to a specific channel, synchronizing with it, and decoding incoming signals.

Why is the similar ability of our nervous system so bewildering? Why is there still no clear understanding of how this happens despite the huge accumulated empirical material on anatomy and physiology? The answer, perhaps, is such: because the question of what representations are, whether they exist at all and if they do, how they are formed and transmitted, remains open. The physical and technological aspects are still an unresolved issue.

Communication channels in the brain converge as we move from central processing units to peripherals and back. At the ends of these channels, there is a huge, at first glance, even redundant network of elements and a vast network diverging towards them. Many neurons control the same effectors, and the effectors themselves can perform different movements, and various effectors can perform the same actions. This is not like a puppet hanging on strings that run parallel and connect the higher control directly to each lower movement. How can all this work so well (normally) with all the paradoxes of such a spatial solution and time constraints?

So, there is morphological convergence in the motor pathways between the cortex and the periphery. This is all analogous to the architecture of sensory pathways. But despite all this merger, excellent selectivity remains. A mechanism is needed to preserve the necessary and sufficient level of selectivity at the required level of generalization. This mystery is insoluble within the framework of the classical paradigm, which we will call a "puppet on strings" model.

One of the active participants in the BMI process, the neuroscientist Apostolos Georgopoulos, showed in the 1980s that the activity of even a small group of neurons in the motor cortex is connected by a cause-and-effect chain with limb movement. This served as a start for developing technologies for the connection between the recording of such an activity and the reconstruction of limb movement.

The author wrote: "A basic finding in motor neurophysiology has been the discovery of directional tuning in space (Georgopoulos et al., 1982), namely the orderly variation of single cell activity with the direction of arm movement, such that activity is highest for a particular movement direction (the cell's "preferred direction") and decreases progressively with movements farther and farther away from the preferred direction … As is the case with receptive fields being interconnected across sensory areas, an approximate topographic correspondence would interconnect directional tuning fields across various motor areas … If we were to trace the sequence of events across brain areas following the onset of a visual stimulus instructing a movement in that direction, we would be impressed by the close temporal and directionally tuned engagement of the various areas, from the onset of the instructing stimulus to the onset of the movement. Although it is reasonable to assume a progression of directional information transmission from posterior (visual) to anterior (motor) areas, and, therefrom, bidirectionally to subcortical (thalamic, basal ganglia, and cerebellar) loops, it is remarkable that drastic changes in cell activity were observed, at the limit, as early as 40ms following stimulus onset in a randomized movement direction task in the motor cortex. These observations point to a strong directionally tuned co-activation among motor areas; we call this directional motor resonance. It is reasonable to suppose that this functional resonance emanates from the underlying pervasive anatomical, topographically organized, connectivity among the various motor areas and leads to the initiation of movement in the intended direction, an essential aspect of motor control. The orderly topographic connectivity constitutes one fundamental aspect of CNS motor control by which various brain areas become directionally aligned, so to speak" (Mahan, Georgopoulos, 2013).

Is it really enough to have orderly topographic connectivity and align all in a row? Transmission within a single synapse takes at least 0.5 ms, and the transmission speed along axons ranges from 0.5 m/s to 120 m/s, depending on myelination. If we take the number of synapses and axons that exist between the visual receptors, all the subcortical, cortical structures and motor effectors, the time must be hundreds or even thousands of milliseconds. This is too slow and does not correspond to the observed speed of signal processing and control of the actions performed by our brain.

The authors use the word resonance, but it does not carry the physical meaning that usually stands behind it — the response of an oscillatory system to external oscillatory action. For them, resonance is about the topographic connection. A well-ordered topographic organization is, of course, needed, and it exists. But is it enough to put everything in a row ("become directionally aligned")? And is there a linear row? Moreover, it is not enough to say that there is some kind of topographic correspondence to answer the question that the authors ask at the beginning of the article. The question of the "puzzling fact that, despite all of this convergence, a remarkable specificity in receptive field size is present." And even to the question about "how the directional tuning arises, i.e., what are the relevant synaptic interactions that underlie the shaping of single cell activity to a typically broad tuning function?" (Ibid).

In this model, there are no answers to the questions that arise in the premise and the questions that arise due to the introduction of terms during the creation of the concept. What is directional tuning, except for the apparent fact of the connection between the activity of neurons in the motor cortex and movements? Perhaps the directional tuning is a representation? But then the question is still the same: how does it arise and work? How can the activity of one cell and even a population of cells of the cortex with a "rough but orderly topographic connectivities among brain areas" lead to such selectivity of commands and responses to them, provided such convergence of pathways? And what is the simultaneous activity? How does it differentiate if it is simultaneous? How does the music of movement arise in this noise of voices of motor zones' neurons? And how can such a topography "align" areas of the brain for such simultaneous activity?

As usual, the emphasis on the spatial aspect does not answer the questions and even multiplies them, creates paradoxes and dead-ends. Maybe this is a situation specific to this particular model and article? Let's take a textbook as a reflection of the mainstream and scroll through Section III, "Movement and Its Central Control" (Neuroscience, Fifth Edition, Purves et al., 2012).

At the very beginning, the authors state: "Movements, whether voluntary or involuntary, are produced by spatial and temporal patterns of muscular contractions orchestrated by neural circuits in the brain and spinal cord. Analysis of these circuits is fundamental to an understanding of both normal behavior and the etiology of a variety of neurological disorders" What a great start. But turning the pages, we come to a disappointment: "Despite much effort, the sequence of events that leads from thought and emotion to movement is still poorly understood" (Ibid).

The authors then move on to the lower levels of motor neurons associated directly with muscle fibers. It is a detailed anatomical and topographical description, but it does not answer the question in the title about central control. But we have no other control. As the authors themselves wrote at the beginning of the section: "Most upper motor neurons, regardless of their source, influence the generation of movements by modulating the activity of the local circuits" (Ibid).

How do they modulate? How do they influence? And what about the high-level representations of the movement? The answer: "poorly understood." Then again comes a detailed description of the anatomy and block diagrams with division into subsystems and arrows of their interaction. There is a more or less detailed description of subsystems: motor cortex, brain stem, cerebellum, basal ganglia, and local motor neurons. It is interesting and informative but does not even pretend to answer basic questions. Both the metaphor of the orchestra and the reference to spatial and temporal patterns, which was at the very beginning of the chapter, are buried in anatomy and linear schemes, and the sequence of events is not clarified. Perhaps because the authors consider it a metaphor and not an analogy of the physics of the process? And then everything is poorly understood, despite the great efforts and the flow of anatomical details. We again vividly see how ignoring physics and technology and bias on physiology alone leads to a dead-end.

And one more thing: when it comes to temporal issues, the authors are strikingly laconic: "Much of the spatial coordination and timing of muscle activation required for complex rhythmic movements such as locomotion are provided by specialized local circuits called central pattern generators" (Ibid).

It turns out that the entire organization of complex movements is reduced to local pattern generators. It seems that the issues of temporal and spatial coordination of the work of all levels of the system are not central themes or have long been resolved and are so well known to everyone that they are not even worth mentioning in the textbook. Maybe they are already known even to children in kindergarten? Then why is the sequence of events from representation to action poorly understood?

In the section describing the higher levels of movement control, detailed information on the functional anatomy of the corresponding zones and the body map in the motor cortex is given. This map has been studied for more than a century. The most important observation was at odds with what many neuroscientists expected. With the development of the degree of resolution of the technology of electrical stimulation, it was found that even the smallest currents caused the activation of several muscles and the simultaneous inhibition of others. This indicated that in the cortex, not individual muscles are represented, but the movements themselves. Moreover, within large areas (for example, an arm or a face), a particular movement could be caused by stimulation of entirely different points. Besides, various movements could be caused by stimulation of the same zone.

It would seem that this should shake the foundations of the "puppet on strings" model. But from the textbook authors' point of view, this indicated that neighboring zones are connected locally and with the spinal cord. They are connected, but is this the answer to why there is both selectivity and generalization in the same place? Rather, how is this selectivity and generalization technologically organized? If everything is connected, then this gives an answer to the combination problem, but does not help in any way with the selectivity problem, does not solve the paradox of convergence while maintaining differentiation.

Why, as soon as the researchers triumph, having reported the finding of "finger neurons," it turns out that other neurons in a completely different place can move a finger, and these same neurons can move not only a finger? The story repeated the studies and models of the sensory zones functioning. When they find a "grandmother neuron," it soon turns out to be a "grandfather neuron." The reason is simple: the localizationist paradigm is fundamentally wrong.

Who said that locally focused electrical stimulation leads to the local response? Many have said so. But are they right? Just imagine that the electrode is a stone that you throw into the water. If it is large (like old electrodes), then there will be a significant disturbance and one pattern; if small, then another. Water molecules remain the same but participate in completely different patterns. Imagine that you are throwing a stone not into stagnant water but into a rough sea, where there are already many wave patterns. Your stone will, of course, trigger its pattern, but it will naturally flow into existing ones. The interference pattern will change, but at a distance, it will be challenging to make out what is happening and where the pattern of your stone is. Neurons are oscillators that take part in wave processes in an environment of the brain (by the way, mostly consisting of water). The influence of the electrode is a change in these processes, the intrusion of an external source of oscillations, which are accepted by the system and absorbed by it because they are related to it by the physics of the process.

The system never stands still. There are always waves in it. Just like the neuronal signal, the electrode signal is not "pulling strings," but only part of the process of forming wave patterns of representations. That is why the electrode stimulation of the same neuron and population can lead to different patterns, and the effect on different ones can lead to the same patterns. On the one hand, these patterns as representations of movement, external and internal signals are individual. On the other, they can be formed by entirely different elements tuned to the parameters of this pattern. And this is not a "topographic correspondence" as in the Georgopoulos model, but a fine regulation of oscillatory processes, their propagation and interaction. In this sense, the motor cortex, and indeed the motor system as a whole, is no different from sensory systems. A living system uses the same physical mechanisms and technologies for all representations. The principle is the same; details and levels differ.

What is, for example, a directional tuning in this light? It is the creation of a representation of the movement direction as a wave pattern, which is projected onto the effectors and correlates with the introjection from them, as feedback in the form of wave patterns. It is also the answer to the question of how it arises and how it works. We need to look at the physics of the signal transduction and synchronization process. It is also the answer to the convergence/selectivity conundrum. Different patterns can go by the same channel at the same time. Receivers need to be tuned to this waveform. Many receivers can reproduce the same waveform. Various transmitters can generate the same waves, and one transmitter can generate different waves. This process can take place in both parallel and serial communication channels.

The point is in the model. Under one paradigm, the mechanism of work and sequence of events, i.e., the technology, is poorly understood; under the other, it becomes clearer.

It is not enough to say that "the representation of muscle movement is not organized at the level of individual muscles or body parts" and that "it suggests a dynamic and flexible means for encoding higher order movement parameters that entails the coordinated activation of multiple muscle groups across several joints to perform behaviorally useful actions" (Ibid). The authors of the textbook simply bypass the question of the mechanism of this dynamic and flexible coordination. They just state "the fact that the same site in the primary motor cortex can encode different trajectories of motion depending upon the starting position of the limb suggests that multiple parameters of movement may be selected by the relevant ensemble of upper motor neurons to achieve a behaviorally useful action" (Ibid).

But how does this ensemble choose the parameters? How does one zone control various movements of the same effectors and different effectors? "What is actually represented in the motor cortex is the intention of movement, rather than movement per se" (Ibid). But the intention is a representation, a model of movement. The same questions arise. What is this representation physically? How is it encoded?

When the authors use verbs such as "plan," "initiate," "compare," "correct" in relation to the activity of brain zones, the question inevitably arises: are there homunculi sitting in each zone and performing all those functions? Because without answering the question about technologies of encoding representations and the physical mechanisms of their creation, differentiation, integration and coordination, saying that an ensemble of neurons selects parameters of movement is the same as saying that a homunculus in the head is selecting them.

The authors note that "unraveling the details of what is represented in motor maps still holds the key to understanding how patterns of activity in the primate motor cortex generate the rich repertoire of volitional movement" (Ibid). But we need to add that it is impossible to unravel the details of how movements are represented without a general model of how the brain creates representations. We can go into the details of the motor cortex down anatomy down to the molecular level, but a strictly physiological approach will not open the gates of the Mind. Three keys are needed: physical, physiological and technological.

Let's go back almost a hundred years. Perhaps there were models that, even if they did not make everything immediately understandable (are there any at all?), at least showed light at the end of the tunnel?

At the beginning of the twentieth century, a line of research emerged that was called biomechanics. Physiologist Nikolai Bernstein is considered one of the founders. His works "General biomechanics" (1926), "Studies on the biodynamics of locomotions" (1935), "On the construction of movements" (1947), "Essays on physiology of movements and activity physiology" (1966) formed the basis of modern biomechanics as a science. He studied movements in norm and pathology using the newest and original methods and formulated ideas that were revolutionary for those times.

Of course, Bernstein was not alone and not the first one. Aristotle wrote about the laws of movement of animals and humans. The first book on biomechanics was Giovanni Borelli's "On the movement of animals" (Borelli, 1680). Many physiologists of the 19th and 20th centuries studied movements. But Bernstein created a direction of scientific research distinguished by a combination of functional, physical, physiological and technological approaches. Moreover, he believed that the study of movement was the key to understanding the patterns of brain function. The very formulation of the problem was revolutionary. At that time, the concepts of psychology either in principle excluded consciousness from research (behaviorism) or limited themselves to interpretations of the manifestations of the psyche but did not in any way connect them with the physical state of the substrate (psychodynamic theories). In general, physiology and psychology went separate ways. In both areas, the leading theories did not reveal the physics and technologies of internal processes.

Bernstein realized that the central nervous system initiates, controls, and corrects any movement in constant connection with sensory and motor effectors. The fundamental concept "sensory correction" introduced by him in 1928 was called "feedback" twenty years later by the founder of cybernetics, Norbert Wiener. Bernstein developed the idea that the brain is not a passive reflective "mirror" but an active governing authority that controls the outcome based on feedback. The traditional scheme "signal-reception" or "stimulus-response" was not just supplemented in his theory with a chain of control but completely turned upside down in the literal sense: the head turned out to be the head for everything. This seems to be a simple and obvious statement, but in order to understand all its complexity, it is necessary to imagine the atmosphere in the science of those times. Those who spoke of this point of view were loners outside the mainstream.

The predominant paradigm was the reflex principle put forward by René Descartes in the 17th century. In the 19th century, physiologist Ivan Sechenov created a theory about the completely reflexive nature of the activity of the higher parts of the brain. It was, of course, a breakthrough in science since it introduced the physiological approach to the analysis of mental processes. But it also left a heavy legacy of the linear, mechanistic approach, which is still not overcome.

Ivan Pavlov developed Sechenov's reflex theory and by the time of Bernstein's first publications was already a world celebrity, Nobel laureate and indisputable authority. By the way, there was a bust of Descartes in front of the entrance to Pavlov's office. As heir to a centuries-old tradition, Pavlov did not accept Bernstein's ideas, and their development, of course, required great courage from the young scientist. The typical situation in science: breaking out of the mainstream and creating new directions requires not only talent but also independence of character.

Bernstein attempted to answer the question: how does a living system solve the "centipede problem" by controlling a massive number of motor effectors so quickly and efficiently that it gives the impression of reflex "automatism"? The parable of a centipede that cannot move if it thinks about movement, on the one hand, is accurate since movements really should be faster than thoughts about them, but, on the other hand, it is incorrect, since any action, even the quickest and involuntary, is controlled by the central processor. The difference between the voluntary and the involuntary in this sense is only in the involved levels of this processor. The thought processes in the higher integrators are too slow and energy-intensive to be used during regular movements. Only mastering the new requires the use of higher levels.

Organisms more complex than a centipede have a completely different order of effectors' quantity: there are not hundreds, but thousands of them. For example, an elephant has about 40,000 muscles only in its trunk. How does a living system combine multiple parameters and reduce the degrees of freedom to a state of controllability? All the links in the motor chain have a range of possible movements, which is difficult to calculate even with the help of a modern computer. Can a linear communication system ("puppet on strings") control such a process? The question requires a specific technological answer. If the linear scheme seems to be unrealistic, then what is the actual algorithm of the process?

The fact that the head is at the head of the process seems to be a trivial and even tautological statement. But having said "A," one must also say "B," i.e., show how it manages such a multifactorial process. To say that the brain is in charge is not enough. The question arises: how? And there are so many further questions that it's easier to say, "poorly understood" or even "somehow." The behaviorists did just that: they removed the Mind from the scheme. No Mind, never mind the problem. There is stimulus-reaction, and the rest is a "black box" inaccessible for research and knowledge.

Bernstein believed that the brain makes "probabilistic prediction based on the current perceived situation," "extrapolation for a certain time ahead," creates an advanced predictive "model of the required future" (Bernstein, 1962). But now, you rarely find references to the ideas of this scientist in modern Bayesian models, which have become mainstream and say the same thing.

He also wrote that the goal is the basis of any action, no matter how automatic and involuntary it may seem. He said that in the reflex arc concept leading at that time, a person is presented as a "highly organized reactive machine." However, even at the intuitive level of the everyday understanding of the work of consciousness, it is not correct. It turned out that the leading theories clearly contradicted reality.

Bernstein wrote: "The body does not just react to a situation or a signal-significant element isolated from it but is faced with a dynamically changeable situation, and therefore puts it in front of the need for a probabilistic forecast and then a choice. Allowing oneself a metaphor, we can say that the organism is constantly playing a game with its surrounding nature — a game whose rules are not defined, and the moves "conceived" by the counterpart are unknown. This particular feature of the existing relations — the assessment is not more accurate than with a certain degree of probability, and the active choice of action that overcomes the situation not conditioned by its command signal — this is what essentially distinguishes a living organism from a reactive machine of any degree of accuracy and complexity" (Bernstein, 1962).

In another work, he wrote: "How exactly, by what physiological pathways the image of a foreseeable or required effect of an action can function as a leading determinant of the motor composition of an action and a program of sending a setting element is a question to which any concrete and the reasonable answer has not yet begun to be outlined ... Using the concept belonging to the field of psychology of an image or representation of the result of an action to characterize the leading link of the motor act, with emphasis on the fact that we do not yet know how to name at the moment the physiological mechanism underlying it, can in no way mean non-recognition of the existence of this latter or excluding it from our attention ... However, ignoramus does not mean ignorabimus" (if we do not know, it does not mean that we will never know)" (Bernstein, 1966).

Bernstein, basing on his research over the decades, comes to an unambiguous conclusion: the control of wide variability of movements cannot be carried out without the existence of this movement's representation. And the fact that at that time, there were no hypotheses about the physiological and physical mechanisms of its creation did not mean that this issue should be "swept under the carpet."

Bernstein criticized the reflex arc theory and said that any, even the simplest movement is controlled based on a cycle of feedforward and feedback links between the center and the periphery. He wrote: "In physiology, the great versatility of such a circular scheme of regulation with the help of feedback is becoming more vividly revealed. In a number of functions, where, for a less in-depth look of the former physiologists, the reaction of the organism was exhausted and seemed a single reflex cut off at the end of an open reflex arc, a new, more precise and intent approach reveals an uninterrupted ... circular control process" (Bernstein, 1962).

One of the pioneers of neurophysiology, Nobel laureate Charles Sherrington, warned in his 1906 lectures: "A simple reflex is probably a purely abstract conception, because all parts of the nervous system are connected together and no part of it is probably ever capable of reaction without affecting and being affected by various other parts, and it is a system certainly never absolutely at rest. But the simple reflex is a convenient, if not a probable, action" (Sherrington, 1906).

The term "reflex arc," introduced in 1850 by Marshall Hall, does not reflect the technology of the process and can be deposited in the archives of the history of science. Unfortunately, it is still widely used in physiology. The reason is simple: there is no model that could explain how the brain can control all the movements so quickly.

The article in Wikipedia states: "A reflex arc is a neural pathway that controls a reflex. In vertebrates, most sensory neurons do not pass directly into the brain, but synapse n the spinal cord. This allows for faster reflex actions to occur by activating spinal motor neurons without the delay of routing signals through the brain" (Wikipedia "Reflex arc"). The physiological fact that sensory neurons do not stretch directly to the brain and synapse at the local level is taken for a technological fact that the circuit is locally controlled. It is a mistake.

As Karl Pribram wrote: "A revision of the reflex arc concept becomes necessary because of data not available to Sherrington. These data show that all of the organism's input mechanisms are directly controlled by the central nervous system ... The ubiquitous presence of the central control over receptor function makes almost useless the reflex arc, stimulus-response conception of neurobehavioral organization, let alone psychological function" (Pribram, 1971).

The old concepts were the result of a lack of empirical knowledge about the operation of the system. The apparent simplicity of the "reflex reactions" was an illusion. Researchers could not believe that the system could manage to operate at such a speed of information processing. It seemed to them that only a simple linear and short chain was possible. The paradox is that if the system worked the way the supporters of the stimulus-response paradigm saw it, it would definitely not have time. But the system works proactively due to the projection of representations.

Bernstein believed that the brain creates an image of the expected movement. It is not only about mental images like "I'll raise my hand now." We know perfectly well that our hand can go up without any thought about it. If thoughts do arise, then only after the move. They may happen before action, but they may not occur at all. The verbal level can give commands-representations, but they still have to be synchronized with patterns-representations in sensory-motor level.

Back in the 1930s, Bernstein analyzed many actions, both simple involuntary and complex voluntary, with the help of cinematic recording. Signals from different limbs and joints were recorded. The subjects put on black suits with white marks, and the camera recorded the movements of the marks. In other words, the wave movements of the body were discretized, and a pattern of movements was obtained that could be analyzed mathematically.

Long before modern computational neuroscience methods, Bernstein applied the analog-discrete conversion and Fourier method to body movements. As a result, Bernstein found out that any motion, as a continual event, can be represented as a series of oscillations and predicted to within millimeters as an integral of several harmonic oscillations. If you make a Fourier-type analysis of the movement, it decomposes into harmonics, and if you make a Fourier synthesis, it will fold again as a single wave. But the most important thing is that having a general idea of the motion spectrogram, it is possible to predict its trajectory.

Bernstein wrote: "First of all, one must turn to the fact of the integrity of movement, its unity and the mutual conditioning of its parts in space and time. The depiction of rhythmic movement established by me in the form of a three-four-term trigonometric sum undoubtedly proves the existence of such integrity in time, and this integrity is by no means peripheral, not mechanical, but undoubtedly of central nervous origin. It proves to us that in the central nervous system there are exact formulas of movements (Bewegungsformeln) or engrams of the latter, and these formulas or engrams cover in one of the brain's circuits the full process of movement throughout its entire time course ... This makes one think, still in the form of a hypothesis, but very persistently asking for itself, that the area of localization of these supreme motor engrams also has a topological ordering according to the type of external space or motor field (at least not according to the type of muscular-joint apparatus)" (Bernstein, 1966).

Modern research confirms this assumption. For example, in experiments with monkeys, the researchers stimulated the motor cortex with electrodes. They found that in one case stimulation caused the hand to close, move to the mouth, and the mouth to open, and in the other, it caused the hand to open, rotate until the grip faced outward, and the arm to project out as if the animal were reaching (Graziano et al., 2002).

The authors of another study of the monkey motor cortex tried to investigate if the old hypothesis about direct muscle control has its foundation in the actual workings of the brain (Griffin et al., 2015). The experiment showed that only 20-30% of the neuron activity coincided with the muscles movement or was within

±45°. The same number had the opposite direction or differed by ±135°. The rest were somewhere in the middle, from ±46° to ±134°. To put it short: there is no directional tuning.

The authors tracked the activity of individual neurons. The same neuron actively worked when the monkey moved its arm, wrist and finger. But the directions of movement of these parts of the body were different in relation to the target. The authors took other neurons and everywhere the same thing: extreme mismatch. The neuron "pulls" different strings, the effectors move in various directions, there is no connection between the preferred direction of the neuron and the direction of movement of the body. Maybe there is simply no preferred direction and localization according to the principle of topographic correspondence?

The authors of the preferred direction hypothesis suggested: "The most likely explanation lies in the pervasive, albeit rough, topographical correspondence in anatomical connectivity and in the parallel presence of seemingly non-specific inhibitory mechanisms. What saves the day is the pervasive congruence among the various "directions" and the covariation of multijoint limb kinetics and muscular activity with movement direction" (Mahan, Georgopoulos, 2013).

The reality is that there is no topographic correspondence, even rough one. There is no pervasive congruence between neuronal "directions" and effector kinetics. Instead, there is an extreme disparity. It turns out that nothing saves the day for the model because the kinetics and directions of movements are there, but there seem to be no directional tunings of neurons. But the monkey is moving in the right direction. It is a paradox, even a dead-end, not for the monkey, but for the "puppet on strings" model. The mismatch is not in neurons and muscles but in the theories and the actual phenomena.

In one fell swoop, the experiment gave a negative result both for the hypothesis of neuron specialization in "pulling strings" of specific muscles and the idea of specialization in the direction of movement. But what about the previous experiments on which Georgopoulos built his model? They showed that the activity of even a small group of neurons in the motor cortex predicts limb movement, i.e., is connected with it by a chain of cause and effect. The chain certainly exists. It is foolish to deny the fact that neurons control body movements. The point, as always, is in the hypothesis as the answer to the question. And this question is how neurons encode motion.

Suppose the answer is that neurons encode actions of effectors with average activity. In that case, the presence of a superficial relation between average activity and action serves as the basis for "a basic finding in motor neurophysiology." But if we go deeper, it becomes clear that the discovery turns out to be a dead-end direction.

There is neither topographic correspondence nor direct relation between neurons in the motor cortex and specific actions. The same neurons can be associated with an agonist in one movement and with an antagonist in another. Some neurons are active when the wrist squeezes firmly, while others are active when weakly, although the same muscles are involved. It turns out that there is no specialization either in directions or in muscles. There must be some other option for motion control technology.

What conclusions do the authors draw? "The key result of the present study is that for many CM cells there is a major disparity between the cell's preferred direction and the preferred directions of its target muscles. We interpret this result as indication that individual CM cells are functionally tuned ... From this perspective, the multiple functions of the target muscles are represented by the activity of separate populations of CM cells ... A major question raised by our results concerns the origin of the functional tuning that we observed. It is possible that the functional tuning reflects an explicit representation of different motor functions" (Griffin et al., 2015).

But it is not enough to say that neurons are tuned for a representation of movement and not for direct muscle or joint control. Of course, it is a change from the unrealistic "puppet on strings" model. But we need to elucidate the physical mechanism and technology that stand behind such tuning. A model should explain the combination of the observed complexity of movements and the speed of their control.

Let's start with speed and look at a cheetah hunting a gazelle. Decisions about the movement before the attack are made strategically: the hunter carefully prepares for the attack, evaluates all the parameters of the victim, the situation, the position, and state of the body. Here is the topology of space, the direction of movement, speed, acceleration, effort, and possible changes, the dynamics of these factors in the future. There is time before the attack. But when the attack begins, time almost disappears. The cheetah is the fastest ground predator: acceleration to 100 km/h in 3 seconds. It accelerates faster than a jet plane. When we look at the recording of an attack at normal speed, our visual system simply does not have time to register the subtleties of the dynamics of the situation. You must slow down the recording to see the details.

Moreover, we do not have time to think about the actions of the cheetah. If we start to think about what the cheetah needs to do at each point of the changing situation, we will have to slow down the recording even more, up to a complete inconsistency with reality. But the reality is this: the cheetah continues to move very quickly and purposefully along a complex trajectory and react to the slightest changes in the trajectory of the victim. Does it think about its actions? Of course, otherwise, it would not exist. Its brain controls the body. But action decisions are made instantly (milliseconds).

Usually in films about animals, the narrator says: instinct prompts, instinct controls, instinct leads, etc. One gets the impression that there is a certain entity that sits inside the animal and controls the whole process, and the animal is just an automaton acting according to the program. However, it is not so. It is not an entity named Instinct that controls the process, but the process we call the Mind, which occurs in the cheetah's brain, controls the entire body. The cheetah thinks about what it is doing.

But how can it think so fast? The question is, what do we mean by the word "think." If we imagine this as internal verbal processes (thoughts), we are stumped because we reduce the whole process to a small part. We, of course, are not as fast in movements as cheetahs, but in general, our actions are much faster than thinking about them. Verbal representations, as a superstructure over sensory-motor representations, are very slow. They have a slightly different function than directing a vast number of body movements. But representations at the sensory-motor level of consciousness are very fast. Otherwise, we would not have survived.

But let's not forget about the gazelle. It wants to live too. The cheetah is considered the most effective predator among African cats, but its attacks are successful in one of two trials. Before an attack, a predator can make probabilistic assumptions, project a model of reality for a long time. But during an attack, the situation in time is radically different. But this does not mean that there is no projection, that there are no assumptions and decisions. If you think that everything happens "automatically," then you either have never seen this spectacle or looked inattentively.

If in doubt, scroll in slow motion: although the dynamics of the situation have its own tendencies, it shows a complex aperiodic pattern that cannot be fully predicted in advance. Like any such phenomenon, it can only be predicted probabilistically. That is why not every attack is successful, although the cheetah and gazelle's speeds are comparable. Of course, the cheetah's reality model has proven options. That is why it is a model. But it is not an automatism in the sense of the lack of variability of solutions following the current dynamics of the parameters of the environment and the body.

During an attack, when there is almost literally no time, the number of parameters does not fundamentally change, and decisions must be made here and now since life depends on them. But there is some reduction in the degrees of freedom. For example, the victim is singular and not a whole herd of potential victims. But the parameters of directions, efforts, accelerations during the attack remain numerous. Moreover, their variability is increasing; the dynamics are growing. In milliseconds, the nervous system must process many signals and produce one correct answer. Because the other will be wrong, as it will lead to an error in the form of an unsuccessful attack. If the cheetah is exhausted before the attack, then it can be the last for it.

The life of all cells of this unicellular society called "cheetah" depends on the message of the brain cells. The responsibility is great, but its delegation took place a long time ago, and there is no turning back. What should the neurons do in a situation of such responsibility? Send numerous discrete impulses to numerous muscle fibers and joints ("pull strings")? There are two aspects that make this scheme unrealistic: time and space.

Let's look at the time constraints of real life. We know that cheetahs and gazelles are fast animals. But they are not record holders. The chameleon's movements are almost comically slow, but its tongue accelerates to 100 km/h in a hundredth of a second when attacking. The mantis shrimp makes a filigree hitting movement of its claw in 3 milliseconds. But this is not a record yet. Insects' speeds are sometimes so high that even a special camera captures only a part of the details.

What we hear as a continuous undifferentiated buzz is complex wing movements with capabilities so far unattainable for our artificial flying technologies. What on the surface looks like an "automatic" reaction, when examined in detail, are the most complex trajectories of movements and, consequently, computational processes and the transfer of information capable of creating these movements.

The speed of the ant's jaw is 64 meters per second (230 km/h). This is the fastest movement in the animal kingdom that we have studied. The quickest human movement (blinking) occurs two thousand times slower. To capture the ant's movement, we need a camera with a frame rate of 50,000 per second. For us to see this movement, it has to be slowed down 1667 times. This movement has more than one goal. The ant both bites and jumps with the help of its jaws. Jaws, like a catapult, throw it 39.6 cm in 0.27 seconds. The ant must decide what it is doing: either it bites or it jumps. But if it has already decided something with its brain, then the same brain instantly controls the purposeful movement of the body. Jaw clenching happens in 0.00013 of a second! Here is an absolute record.

Of course, the small distance between the neurons of the ant's brain and the effectors of movement plays a role here. Human neural pathways are much longer but fast enough to provide almost instant actions. The conduction speed of different paths is estimated in the range from 10 to 120 m/s. We cannot compare in speed with the desert ant, which can run at 50 steps per second or 660 km/h. It is another record in the animal world. But regardless of who we give the name of the record holder, there is a fact of the implementation of body movements under the control of the nervous system during a time frame when it is impossible to speak about any variability of the average firing rate. Even if we take relatively slow movements, they also occur within a few spikes. There is no way to encode a complex pattern with an average speed of 2-3 identical shots.

We have discussed in detail how the Symphonic Neural Code (SNC) hypothesis within TTT leads out of the dead-ends of the firing rate paradigm when it comes to the question of the informational density of the code within the time constraints of real life (see "Part Four. Algorithm of the Mind"). Here we are interested in how complex information is transmitted to ensure the speed and accuracy of movements.

The mainstream neuroscience models describe the transmission of information in the brain as "spike trains." The fact of life is that in the brain, the "train" just started moving in one place and has already arrived at another. As we have shown, this amazing speed cannot be explained within the paradigm that takes neural activity for discrete spikes transmitted along the wires. Neural code is not purely digital and neural channels are no wires.

TTT proceeds from the assumption about the wave nature of information transfer. It deals with the physics of the neural code and the physics of the medium through which encoded patterns propagate. The secret of the almost instantaneous exchange of information between the system elements, despite branching of the network, long distances (compared to the size of cells), and the limitation of the physical capabilities of each cell, is in the physics of waves. They can have structured patterns, as carriers of information, and are the means of its fast and efficient transmission without movement of the substrate. Waves propagate in any medium, at any distance, can have any speed, and retain the pattern structure even in the presence of obstacles.

When passing from one medium to another, a distortion of the pattern and a change in speed, attenuation resulting from energy dissipation, is possible. To minimize these effects, the system should try to maintain a homogeneous environment. The environment of the nervous system, with all its functional and anatomical diversity, is relatively homogeneous: from a physical point of view, it is an aqueous solution. As we have discussed earlier, the system also uses various technologies for overcoming channel limitations and guiding the waves.

The wave nature of representations also allows for solving problems related to the spatial aspect. Let's get back to the riddle of neural pathways divergence and convergence. Why does the brain have such a "mixed-up" structure? The answer lies in the technological aspect. From the point of view of organizing communication channels, the brain uses a component (different streams by various channels) and composite (different streams on the same channel) solution. Primary signal converters represent a component aspect: different channels process and transmit various parameters of a signal. The component solution at the input allows differentiating and avoiding crosstalk. In neuroscience, it is called "labeled lines." There is a transition to composite solutions at the subsequent stages when the encoded parameters are transmitted over common channels. This simplifies the connection between the modules of the system, which is highly complex anyway due to the vast number of network elements. The composite version saves space, time and energy during transmission. But it means that the brain has to deal with the problem of the transmission of multiple data streams over one physical communication channel. This requires multiplexing technologies.

Let's turn to the analogy with artificial technologies. Here are the main ways of multiplexing:

1. Frequency Division Multiplexing (FDM) is about separating a common channel into frequency sectors (for example, radio channels).

2. Wavelength Division Multiplexing (WDM) involves transmission over a single channel at different wavelengths. The technology is based on the property of waves with different lengths to propagate independently of each other.

3. Time-division multiplexing (TDM) involves dividing data into time slots that are separate for each channel and sending these "frames" in a specific order.

4. On-Demand Multiplexing (ODM) where the total output stream is formed by the incoming channels, through which data blocks (packets) arrive at different time intervals.

Do you think the brain uses FDM and WDM types of multiplexing technology? Of course. Moreover, it uses waves with various frequencies and lengths for different purposes. We will be dealing with the functional roles of these waves in the next part of the study. Here we just illustrate the principle: specific sets of information can be transmitted by waves with different parameters that can exist simultaneously in the same medium and easily "walk" along one channel without mixing.

TDM and ODM are about creating the rhythm of the process. It is like a timetable for the arrival and departure of aircraft. If you arrive on time, you are welcome. If you are late, circle over the airport for a while. If you managed to fly away on time — well done, if not — wait for a free runway. Ideally, everyone arrives and leaves on time. Waves with different phase portraits can create rhythmic scheduling.

Once again, the physics of waves allows to solve the technological problems facing the brain. Now let's consider the control of the complexity of movements. Will the hypothesis of wave representations help in this matter? For this, we need to remember the basic physics: waves are the propagation of oscillations; interaction oscillators tend to synchronize.

Bernstein wrote: "The coordination of movements is overcoming the excessive degrees of freedom of a moving organ, in other words, its transformation into a controlled system" (Bernstein, 1966). One of the modern followers of Bernstein, the neuroscientist Rodolfo Llinas, in his book "I of the Vortex: From Neurons to Self," describes the huge number of degrees of freedom that the system faces when making even a simple movement (Llinas, 2001).

Each muscle is made up of hundreds of motor elements (fibers). The product of the number of such elements by the number of muscles participating in one simple movement is the number of degrees of freedom of a given movement, and it is enormous. Llinas notes that even to get milk out of the fridge, we have to use about 50 muscles, which gives about 10^{15} combinations of possible muscle contractions. It is a cosmic number. If the brain were to evaluate each and choose, it would have to make 10^{18} decisions every second. That would mean its speed as a 1 million gigahertz processor. The brain is much slower than modern processors in computers, nothing to say of fantastic millions of gigahertz. Therefore, something is wrong with the hypothesis.

Llinas writes that an alternative would be that each muscle is independently controlled and that the motor system is a set of "parallel processors with one for each muscle" (Ibid). As we have mentioned earlier this hypothesis does not comply with reality too. He asks the questions that Bernstein asked many decades ago: how can you solve the problem of controlling a vast number of parameters and reduce their number without losing the quality of consistent and smooth movements? Like Bernstein, he writes about the operation of the system in discrete time intervals, the breakdown of motor tasks into a series of minimal time windows. But then, an integration of such discreteness into continuity with a basic synchronizing pulse is required.

At the beginning of the 20th century, it was discovered that there is a constant muscle pulsation in the frequency range of 8-12 Hz, which was called "physiological tremor." Then it was proved that the beginning of the movement coincides in phase with this tremor. Later studies confirmed that it does not depend on speed, direction, or effort during movements. Moreover, it remains as constantly ticking metronome in a calm state without movement.

In the previous part of the study, we showed that the same frequency range of brain waves acts as a synchronizing basic pulse for the other ranges. We will be dealing with further details of frequency interaction in the next part. Here we just note that low frequencies of muscle pulsation serve the same purpose and the coincidence of the range is not surprising at all.

Maybe the solution to the degrees of freedom problem is not only and not so much in increasing the parallelism of channels and processor speed, but in synchronizing the oscillations? Of course, processor speed, the number of communication channels and their topology are important, and they have improved in evolution. All of these factors work towards one goal: ensuring effective movement.

Hypothesis:

Synchronization of oscillatory processes in the nervous and muscular systems solves both the problem of control and reducing the degrees of freedom for the implementation of accurate and effective movement.

To illustrate, let's return to the centipede problem: how can it manage so many legs? Even a hundred is a management problem, and a real centipede has many more legs (up to several hundred). The limbs of any living system typically move following rhythmically organized patterns, reflecting the synchronization of one order or another. Even living organisms without a nervous system must control their movements. For the bacterium to move to the place of concentration of the nutrient, it must synchronize its flagella with each other. Each flagellum is a complex oscillatory system consisting of microtubules, the work of which must also be organized rhythmically. It is much easier to conduct a synchronized ensemble than the one "at sixes and sevens." The elements combined in ensembles can perform different tasks by one team and the same tasks by different teams.

Even a small number of legs require organization in time-space. When we walk with alternating movements of legs, we do not have to use the verbal level of consciousness and think "one-two-one-two" since the brain synchronizes movements at the sensory-motor level. Only the synchronization of the movement of different people requires an external pacemaker, as, for example, soldiers need marching music with a simple meter or the command of a sergeant "one-two-one-two." Our walking is a very complex combination of trajectories, and each requires synchronization of a massive number of elements of the system. But in the end, it is a movement of two pendulums synchronized in antiphase. The degrees of freedom are minimized.

When a kangaroo jumps on its hind legs so that they simultaneously repel from the ground, its brain creates a considerable number of frequency-phase couplings, but the resulting movement of the limbs occurs in the simplest 1:1 synchronization order (unison). The jumps of the rabbit demonstrate the synchronization of each pair of hind and front legs with each other in phase, but the pairs are synchronized in antiphase. Thus, the four legs are transformed into the same two pendulums in antiphase. The giraffe has a lateral distribution pattern: the front and hind legs on each side move in phase, and these one-sided pairs are in antiphase. Synchronization of the horse's legs while trotting occurs diagonally. The gait of an elephant is characterized by alternating movements of the legs with synchronization of a quarter of a beat. The gazelle jumps, synchronizing all four limbs in phase. The run of a cockroach is an example of the synchronization of six oscillators: three legs (front and back on one side, middle on the other side) synch in phase, and these triangles synch with a half-cycle difference. All these examples are given by the authors of the article "Coupled oscillators and biological synchronization," who emphasize that limbs are not passive mechanical oscillators but complex systems controlled by complex neural ensembles (Strogatz, Stewart, 1993).

What can physically and technologically provide the transmission of a complex information pattern and the action based on this pattern? What can provide speed, efficiency, complexity, coherence and meaningfulness? Physics and technology of the wave process. What ensures an efficient organization of the process? Synchronizing these wave patterns. It is not enough to create a pattern at the level of control elements, although the entire central nervous system is tuned in to this. But is has to ensure that the receiving elements are configured for fast and accurate communication and provide the same fast and accurate feedback.

We must not forget about one more "ace up the sleeve." We are talking about the very essence of the system's operation algorithm, about the PAAL as the constant iterative process of superimposing and comparing projection and introjection waves. The secret of this "trick" is that waves of existing representations converge with the current introjected signals, and as a result, only the difference has to be highlighted. If the system is in a more or less familiar environment, such a scheme allows it to spend practically no time or energy on additional computational processes but immediately transmit the final representation of either a cognitive or motor act. One or two spikes of neurons participating in the wave pattern are enough for the "picture" to come together. Only new signals and new solutions require additional costs.

Here it makes sense to summarize the "tricks" that ensure the speed and efficiency of the technologies of the Mind.

Hypothesis:

To ensure the efficiency and speed of internal information flow, a living system uses the following physical mechanisms, technologies and algorithms:

1. Information richness and flexibility of the hybrid analog-discrete-analog code.

2. The wave nature of representations, providing speed and unity while maintaining the differentiation of flows.

3. Synchronization as a mechanism for the interaction of all system elements for the efficient transfer of energy-information.

4. Iterative PAAL algorithm, providing a "game" ahead of the curve due to the projection of the accumulated stock of representations as a reality model.

The first element creates the basis for everything: if the information is compactly compressed, it can be transferred quickly. The second establishes the foundation for combining local information density into a global one. The third creates the basis for the interaction of elements at any moment. The fourth allows interaction to be proactive, not reactive. For an analogy, we can imagine this as the interaction of a central computer and an external device (periphery, effector).

If the information density of the code is high, then the transmission of a coherent command and receipt of feedback is highly efficient. This is computationally expensive at the stage of code creation. Still, it is justified by the fact that the time and energy costs are significantly reduced when it comes to interaction. If both devices (central and peripheral) work in the same format and the same range, this again reduces the interaction cost. However, initial costs are required for synchronizing devices, bringing their settings to a joint base. They must be ready to interact. As the saying goes: train hard, fight easy.

In addition to co-tuning and a single format, they must have uniform algorithms and software. If the central device and periphery speak different languages, the system will not work. They must have common representations of their tasks. And these representations must be ahead of the current tasks. But they should be updated and corrected. The presence of "homework" is the primary condition for the temporal and spatial efficiency of the result.

It is the same in the nervous system: the commanding, receiving and effecting elements must be part of the playing ensemble, in which there is a constant basic pulse. They must be tuned to the general rhythm, have the appropriate characteristics of the impulse response, but have their own voice as amplitude-frequency characteristics. Moreover, if we remember that not only neurons but also any cells and their populations are oscillators, then we will understand why there is a constant "tremor" in muscle tissues, basic internal activity. And in neurons, there is a basic pulsation both at the level of the cell and at the level of populations.

And along the entire chain from the upper echelons to the lower ones, there are repeaters and the creators of local oscillatory activity. Here it is pertinent to recall such a concept as central pattern generator (CPG). The name itself would have been apt if not for the word "central." It is misleading, as it gives the impression that we are talking about processes in the central elements while the phenomena at the periphery are described. It would be more accurate to call them pattern generation centers. A change of word order has profound implications.

Still, the story about the phenomenon itself is more complicated than the name issue. What will the textbook say? "The contribution of local circuitry to motor control is not, of course, limited to reflexive responses to sensory inputs. Studies of rhythmic movements, such as locomotion and swimming in animal models, have demonstrated that local circuits in the spinal cord, called central pattern generators are fully capable of controlling the timing and coordination of such complex patterns of movement and adjusting them in response to altered circumstances ... A principle that has emerged from studies of central pattern generators is that rhythmic patterns of firing elicit complex motor responses without need of ongoing sensory stimulation" (Purves et al., 2012).

You see how it turns out: there are reflexes and pattern generators in the spinal cord, which by themselves, without a brain and sensory feedback, can control complex patterns of movement. It would seem that this ruins not only the building of all theories trying to answer the question of how the brain works but the whole evolution: why did it try so hard, created such a complex system when reflexes and rhythm generators at the peripheral level are enough for complex motor activity? It is a puzzle.

Let's look at the phrasing in the textbook from a technological point of view. CPG is credited with creating complex patterns of movements and the ability to correct these movements without the need for feedback from the sensors. These are wonders of technology. Simple oscillation creates complex patterns while regulating itself without any external information. Questions arise. How does it create complex patterns? How and, most importantly, why does it change these patterns if there is no information from the outside? In this form, it is technological nonsense, not a wonder of technology.

Every person who has ever led a village life knows that if a chicken head is cut off, it can run for a while. The spinal centers continue to work autonomously for a short time. The chicken will just run around like a brainless chicken, literally. No sensing is required, and the generation of muscle rhythms continues. But such movements can hardly be called purposeful and self-correcting.

In science, the study of such phenomena began a hundred years ago. In 1911, Thomas Brown demonstrated that the cat's spinal cord could create rhythms of movement in the absence of commands from higher centers and afferent connections. These and other observations led to the emergence of the theory of central pattern generators. Brown called them "intrinsic factor," and the modern term appeared in the 1970s (Grillner, Zangger, 1975).

Within this theory many versions of the explanation of the mechanism exist. Including the one expressed by the authors of the textbook: it is an element that creates complex patterns of movements. And it seems that it simply states the fact: if there are movements, there is an element that creates them. Moreover, it has its evolutionary rationale. This is how they usually write: the simplest chordates do not have a brain but produce movements; therefore, CPGs are ancient phylogenetic centers for creating complex patterns.

It is logical, but again, there is a mistake in the premise that a simple rhythm gives rise to intricate patterns. It is a technological contradiction to which the authors of such hypotheses simply do not pay attention. A simple question: how do they imagine it? For example, how can a simple, monotonous metronome beat create music? It can help create it as a pacemaker, but it is not music.

The suspicion arises that the matter is not in the generators of simple rhythms, not in the creators of the basic pulsations, not in the pacemakers. They are, of course, needed as synchronizing elements. But representations of movements, no matter how simple they are, still have to be created by elements capable of generating complex patterns and necessarily having a feedback loop for correction and learning. This is a technologically realistic scheme.

Brown proposed a mechanism for the operation of the CPG, which is called the "half-center oscillator." Its essence lies in the changing activity of flexor and extensor motor neurons. This model was developed in the 1960s and 1980s when a more detailed description of how this might work was given (Jankowska et al. 1967, Lundberg, 1981). Key points of the model: a separate CPG controls each limb; each CPG possesses two groups of activating neurons (half-centers), which regulate the activity of extensor and flexor motor neurons, respectively; mutual inhibitory connections between the half-centers create conditions for the work of only one center at a given moment; an undefined mechanism of "fatigue" creates conditions for changing roles; a phase change occurs when activity falls below a certain level; inhibition of neuronal antagonists is closely related to the activation of agonists.

It is a simple one-level scheme that other authors developed by adding levels and details of possible connections. But even in this form, it is a model of a classical pendulum. In this case, the mechanism itself was only postulated, and the actual details of the phase change regulation remained outside the model. Besides, the observed phenomena, when studied in more detail on the same cats in the 1970s, showed that not just a change in the activation of flexors and extensors occurs, but there is a subtle pattern that demonstrates a complex rhythm of the change in the phases of muscle work.

The simple pendulum model did not correspond to reality. The creators of more complex rhythms were required, and additional modules called "unit burst generators" (UBG) were proposed. But, as the authors of the review article note, "despite the attractiveness of this proposal, the UBG model has not yet provided explicit solutions for the generation of complex motoneuron activity patterns" (McCreaa, Rybak 2008). And it is not surprising because researchers are looking for a complex pattern in a simple pacemaker.

The CPG model was uncritically accepted and has existed for a hundred years, although there were missing details, to say the least. At the most, it simply did not answer the same technological question: how can a simple pacemaker generate complex patterns? But the point is that CPG has been studied as an independent phenomenon. The systematic approach was not in vogue, or rather, it did not exist.

Then a way out of the dilemma was proposed: there is CPG and a "pattern formation layer." In the end, higher control centers were involved in the scheme, and, of course, it could not do without sensors as regulators in the feedback circuit. By the end of the century, everything returned to the beginning, and the mechanism and its place in the general scheme remain in question. And the simplest rhythmic movements, from which the history of the study of CPG began, turned out to be not so simple: they are very adaptively dynamic.

The Wikipedia article states that CPGs "are the source of the tightly-coupled patterns of neural activity that drive rhythmic and stereotyped motor behaviors like walking, swimming, flying, ejaculating, urinating, defecating, breathing, or chewing." Really? Let the author try to fly like a bird and, probably, he/she will learn that it is not a "stereotyped" behavior. Even simple walking has a very complex mechanism of adaptation to the current situation: the nature of the surface, speed, direction, and so on.

The authors of the review article wrote: "One general problem with the original half-center architecture is that it accommodates only a strictly alternating pattern of flexor and extensor activity with all motoneurons divided into these two groups. During locomotion, however, some motoneuron pools display activity during both the flexion and extension phases of the step cycle and there are differences in the onset and offset of activity in individual flexor and extensor pools" (Ibid). A familiar problem for a linear concept, isn't it? The idea that there are neurons of specific muscle activity again does not reflect the observed phenomena. And the authors of the review make a logical conclusion: "Intrinsic rhythmogenic properties of the half-centers may be necessary but not sufficient for the generation of the locomotor rhythm and pattern" (Ibid).

But one of the leading textbooks on neurophysiology, which has gone through five editions, unequivocally states that it is all about reflexes and CPG. And it was written much later than all the studies that showed that such a model has problems. The mainstream doesn't give up so easily. But even the original "proponents of the half-center architecture have clearly recognized the need for additional circuitry and processes to supplement basic CPG operation" (Ibid).

The authors of the review note another problem: "There is, however, a deeper problem with the original half-center organization. A common feature of simple half-center organizations is that the excitatory interneurons generating locomotor rhythm are connected directly to motoneurons. Consequently, any changes in excitability of the half-centers should simultaneously affect both cycle timing and motoneuron activity. This is an obvious disadvantage when independent regulation of motoneuron activity (i.e., muscle force) and step cycle timing is required, e.g., when marching up and down inclined surfaces. This intertwining of cycle timing and motoneuron activation within a single-level CPG architecture has prompted suggestions for a more complex organization" (Ibid).

The problem is the same: the model of linear connections and "pulling strings," albeit alternately. But reality shows complex, though obeying regularities, rhythmic and frequency interactions. Within the TTT framework, the explanation is simple: it is about representations as wave patterns and their synchronization. And regularities are pretty strict: if there are no definite relations, there is no synchronization.

For example, it has been found that motor neuron activity rhythmic bursts often appear after an integer multiple of the pre-deletion cycle period. Deletions have been termed non-resetting since rhythmic movements occur again without a phase shift in the movement cycle. As the authors wrote: "The most reasonable explanation is that the CPG controlling each limb contains a structure that maintains the timing of the cycles during failures of rhythmic motoneuron activity" (Ibid). CPGs are an essential element of the entire system of synchronization of representations.

Hypothesis:

Motor centers of pattern generation, which are commonly called "central pattern generators," are actually peripheral pacemakers in the system of retransmission of wave patterns of motion representations. They perform the function of maintaining the basic synchronizing pulse, independent of phase shifts in other elements of the network, which provides the ability to return to the general rhythm and allows the transmission of complex patterns of wave representation emanating from central integrators for the implementation of movement and its control.

Here is a simple example: you walk and stumble. There is an instant correction of the representation: the pattern works to restore the previous activity. If you fall, it is one pattern. If your movement phases are shifted (stumbling), it is a different representation. In any case, the basic pulse remains in place and allows the new pattern to represent movement correction. It is more technologically logical to say: there are centers for the generation of simple rhythms as the basis for synchronization, and there are subtle mechanisms for regulating the interaction of all complex space-time patterns as representations of movements at the higher levels. This hypothesis differs strongly from the technological nonsense of simple rhythm producing complex behavior.

If we transfer the classical model of the CPG to our example, then the picture will be as follows: you fall and continue to march while lying. This is a logical conclusion from the model, in which a simple rhythm works to generate movements independent of sensory feedback. Such a model works only for wind-up toys: if one falls, it will continue to move, not moving anywhere. It may sound ironic, but the classic CPG model is a model for clockwork toys, not living organisms. The amazing thing is that it still lives. By the way, it is the basis for many attempts in robotics to create movements close to living ones. No wonder it doesn't work.

I had wind-up toys in my childhood. Speaking of chickens, there was also a clockwork chick.

Its iron legs performed simple rhythmic movements based on a drive mechanism that created pendulum-like oscillations. This is a mechanical CPG. The spring winding by the key creates gradually expended kinetic energy. The chick jumped almost as a real one. For such a "march," the system did not need either a brain or a spinal cord. The CPG acted on its own, creating patterns of movement, and it had no feedback either in the form of a mechanical servo mechanism, much less in the form of sensors.

Sounds like the description of the CPG in the textbook? Word by word. However, there is a problem: the chick could not make complex movement patterns and, of course, could not correct them. It jumped almost as a real one, but in this "almost" there is the gap between live and mechanical movement. Now imagine the chick stumbled. A live chick will get up and run on. The chicken mom clucks for order's sake ("be careful, sonny"), but on the whole, it's a matter of everyday life and fixable.

Here is the mechanical chick destiny:

Such a chick, which possesses both a reflex (key as stimulus → jumping as reaction) and a CPG, will lie on its side with legs moving until the energy runs out. If a live chick did this, then it would also run out of energy. The model of movements based on the reflex and CPG, which has existed in science for a hundred years, can, in essence, be described by this photograph. It is not just a scientific impasse but also a dead-end for any living system if it obeyed technologically absurd models.

What is this nostalgic lesson in mechanics about?

Lesson number one: a system without feedback cannot control movements. Artificial systems have long been based on feedback servo drives with the ability to correct and adapt. Otherwise, all our technologies related to movement would not have gone beyond clockwork toys. For billions of years, living systems have been performing actions based on feedforward-feedback schemes.

Lesson number two: a simple pendulum mechanism alone cannot create complex patterns of movement. It can be an essential design element, but intricate patterns require a complex creating, transmitting, controlling and modifying system. Even in mechanical devices, inside which there is a simple generator of cyclic motion, a whole chain of eccentrics, cam mechanisms, other converters and drives are needed to create more complex patterns at the output. To diversify its movements and to avoid dead-ends in life, the chicken needs a more complex mechanism at the periphery and central control of jumps, flights, and so on. All these centers and peripheries should have feedforward-feedback connections.

It is quite obvious that a living system has many levels of motion control. They are fairly well studied anatomically: from muscles and their biomechanics to motor neurons of the first level, which activate these muscles, and interneurons in the further chain of sensory signaling; from the levels of rhythm generators at the periphery to the higher levels of the brain, starting with the stem and ending with the cortex. Channels for exteroceptive, interoceptive, and proprioceptive information are known. Enormous efforts are made to study the system elements, articles about experiments and review articles about the same experiments are written. And what is the result?

"Yet in spite of all the neurophysiological knowledge, there has been relatively little progress in understanding precisely how the neuronal mechanisms combine with the biomechanics to produce the stability, adaptability and grace of animal movement" (Prochazka, Yakovenko, 2007).

Other authors wrote: "The CPG mechanism has inspired the field of robotics, particularly in the development of small autonomous walking robots, from multilegged insect-like robots to humanoids and active prostheses" (Cheron et al., 2012). But current robotics technologies do not create smooth and adaptive movement. These are complex versions of clockwork chicks. That is why the expression "move like a robot" means discrete and awkward movements.

The robotics engineer, Hans Moravek, wrote in 1991: "Today's best computer-controlled robots are like the simpler invertebrates. A thousand-fold increase in computer power in the next decade should make possible machines with reptile-like sensory and motor competence. Properly configured, such robots could do in the physical world what personal computers now do in the world of data — act on our behalf as literal-minded slaves. Growing computer power over the next half-century will allow this reptile stage will be surpassed, in stages producing robots that learn like mammals, model their world like primates and eventually reason like humans" (Moravec, 1991).

He even outlined the approximate milestones of such an evolution: the dumb robot (2000-2010), learning (2010-2020), imagery (2020-2030), reasoning (2030-2040). The actual dates can differ, but the main thing is the direction of the engineer's thought: from a dumb robot to reasoning through self-learning and representations.

Moravec wrote that after the bright hopes of cybernetics and the first wave of AI, a sobering up came. "What a shock! While the pure reasoning programs did their jobs about as well and about as fast as college freshmen, the best robot control programs took hours to find and pick up a few blocks on a table. Often these robots failed completely, giving a performance much worse than a six month old child. This disparity between programs that reason and programs that perceive and act in the real world holds to this day" (Ibid).

Note that he wrote this in 1991. The situation is changing. The time has come for self-learning algorithms. Dumb robots have grown wiser and can operate in the real world. Like a living system, they need a period of learning, but in general, there is serious progress towards orientation in the difficult conditions of a natural environment. The fundamental step was the transition from linear schemes to iterative forward-feedback algorithms with the function of comparison, evaluation and correction.

But algorithm is only part of a deal. There has to be a physical solution that makes it work in fast and efficient way. To this day, engineering solutions are still within the "puppet on strings" paradigm. Robots remain robotic even if we try to make them look as human as possible. It's not about looks. Living systems are very different from each other in appearance. It's also not so much about the technical details of sensory and movement implementation. Living systems have a wide variety of external sense organs and motion effectors. Could it be the physics of representations? The main secret of stability, adaptability, speed and grace of movements of living systems is simple: wave representations produce wave-like movements.

CHAPTER 8

THE UNIVERSAL PROCESSOR

The richest sensory information about the external world is communicated to our brain encoded in the form of impulse chains. How the brain decodes it, is there a specific difference in nerve signals from visual, auditory, tactile and other receptors and what it consists of — these are tasks that the physiology of tomorrow has yet to solve.

Nikolai Bernstein

For a long time, the concept of localization of mental functions prevailed in neurophysiology. The main reason was that only part of the functions suffered when certain areas of the brain were damaged. Indeed, network elements have their specialization. But it's only part of the story.

The founder of neuropsychology, Alexander Luria, wrote about the dynamic localization of functions: "Higher mental functions as complex functional systems cannot be localized in narrow zones of the cerebral cortex or isolated cell groups but must cover complex systems of jointly contributing to the implementation of complex mental processes working zones that can be located in completely different, sometimes far apart from each other parts of the brain" (Luria, 1973). He made a conclusion about the systemic distribution based on the vast clinical experience of observing patients with brain lesions, especially during the Second World War. That is, the experience of previous physiologists was refined and expanded, and the earlier hypotheses about narrow localization turned out to be the result of the limitations of the experience itself.

But the old paradigms do not give up quickly, and for a long time in neurophysiology, the topic of functional dynamism was almost taboo. To illustrate, let's take the story of one discovery that led to the creation of artificial technologies using natural technologies of the brain. There are devices that allow people to see not with their eyes but with other parts of the body, such as the tongue. Even for specialists, this sounds like fantasy, but it is the reality. This technology called Brain Port is based on a discovery made by neuroscientist Paul Bach-y-Rita half a century ago.

The developers from Wicab Inc. emphasize that the device does not replace vision but adds information for users about the size, shape and location of objects (http://www.wicab.com). The tongue, although it is a very sensitive converting filter, is nevertheless specialized in a different range of environmental signals. The technology so far does not restore sight with a wealth of colors and facets. But for a blind person, even limited information is already vision.

The general scheme of the device: a set of 400 electrodes the size of a postage stamp is put on the surface of the tongue; a camera is attached to the glasses; the controller moves the camera and converts the signals from the camera into a pattern of tactile stimulation of the tongue (white pixels — intense stimulation, black — no stimulation, gray — medium, and the possibility of the reverse order if necessary). What do users see? They describe it as moving pictures painted on their tongue with soda bubbles. It takes time for the brain to gain experience with training examples to create appropriate representations and connect them into a general model of reality. Unsurprisingly, the younger the user, the faster the learning process. Advantages of the device: simplicity and no need for surgical intervention. Please note that the apparatus has no direct relation to the brain code. It only converts one signal of the environment into another (light into electricity) and feeds it to the sensors of a living system. The rest is up to the brain: to process signals and make representations from them.

The fact that the modalities of perception have compensatory dynamics is common knowledge. When there is a disruption in the functioning of one modality (for example, vision), others become more sensitive. But there is one very simple and not yet fully adopted idea: modalities not only compensate each other but can also replace each other in the literal sense.

Consider two modalities: visual and tactile. The retina and skin have a flat 2D organization in space. They are screens-catchers of signals, transforming and encoding them within the system into three-dimensional representations. Of course, they have a specialization. The receptors of various modalities differ in energy perception mechanism, but the process of its transformation and coding in the system is the same. Thus, the question arises: can modalities replace each other? For example, regarding vision and tactile sensations: can a blind person see with his skin?

If you type this question in a search engine on the internet, there will be many options of answers: from esotericism with a bias towards a special gift, inner vision, the third eye, communication with other worlds to specific descriptions of the experiments of scientists. It is gratifying that at least some of the links are dedicated to the great neuroscientist who was not evaluated according to his merits during his lifetime.

Long before neuroplasticity became a popular topic, Bach-y-Rita constructed in the 1960s an incredible device for "tactile vision." It consisted of a camera, computer and chair. The camera sent signals to the computer, which transmitted them to a special vibrator on the back of the chair. Four hundred stimulants came into contact with the skin of a person who was blind from birth. They worked as photoreceptors, i.e., they reacted to the dark and light parts differently. These were the years when artificial information processing technologies were still very cumbersome and not very functional. But even then, the blind, after some time, began to perceive objects and the space in front of them through the skin and its mechanoreceptors. The most exciting thing: the subjects continued to perceive normal tactile sensations and differentiate them from visual ones.

This discovery went almost unnoticed, although it was published in Nature magazine in 1969. Of course, the technology was too bulky to be implemented in everyday life. But the main problem was that the idea was too revolutionary for its time. It ran counter to the mainstream of neuroscience, which professed localizationism and a linear signal-reception hierarchy through special channels. But Bach-y-Rita continued his research. Many years later, he invented a surprising way to cure people with the pathology of the vestibular apparatus by installing a plate with stimulants on their tongue. Special sensors on the plate reacted to the position of the body in space and transmitted vibration signals to the tongue, thus connecting the two modalities of perception.

The result exceeded the expectations of the experimenters themselves. The device not only allowed a person, who had previously been unable to stand and move in space without assistance, to restore a sense of balance and acceleration fully, but left a therapeutic effect even after the session. At first, this effect lasted minutes, but after several months of regular use of the device, the first patient was cured entirely and did not need the device. The brain restored the lost function by creating new pathways, new connections for processing information from the vestibular apparatus itself.

Bach-y-Rita was a great experimenter, and his approach was purely technological: "We don't see with our eyes, we see with our brains ... There's nothing special about the optic nerve. The brain doesn't care where the information comes from. Do you need visual input to see? Hell, no. If you respond to light and you perceive, then it's sight ... Anything that can be measured can be transported to the brain. We can get it to the brain, and the brain can learn how to use it" (Bach-y-Rita, 2003).

We are used to thinking that we hear with our ears, see with our eyes, feel with our skin, smell with our nose, feel with our tongue, etc. It's true, but it's part of the truth. All these are just primary filters converting signals into a common format for their subsequent processing by modulators and integrators. We listen with our ears, but we hear with our brain; we touch with our skin, but we feel with our brain.

Bach-y-Rita's experiments in the 1960s were based on a fact discovered by chance, as is often the case in science. But the one who seeks will always find, although sometimes not what he was looking for. Bach-y-Rita and his colleagues, while experimenting with vision in cats, found that neurons in the visual cortex responded not only to visual signals but also to accidental touch on the cat's paw. Someone in the laboratory accidentally stroked the foot of a cat, and here is a discovery. But not everything is so simple in science: the discovery made its way for several decades. Bach-y-Rita was struck by the fact that refuted the prevailing views on the brain work and devoted all his scientific activities to studying this phenomenon and attempts to use it.

Many of his articles were not published, using a very characteristic argument for people (even for scientists): "This cannot be, because it cannot be." But what was considered impossible is the reality because there are no light waves, vibrations of sound waves, mechanical impacts, odors and other types of external signals in the brain. They are converted to representations in a system's universal format. The difference is only in the patterns which encode light, smell, etc.

Only 33 years after the discovery of Bach-y-Rita, research using modern equipment confirmed that tactile stimulation also triggers activity in the visual zone of the cortex. An article about one of the experiments has a title that says it all: "Activation of the visual cortex by electro-tactile stimulation of the tongue of the blind from an early age" (Kupers et al. 2003). Conversely, areas considered as hearing specialists are quite capable of seeing. Experiments on animals have shown that if the optic nerve is connected to the auditory zone, then after a while, the animal sees again. What is the secret of such universality?

Hypothesis:

The brain is a technological device for converting environmental signals, and the neocortex works as a universal processor to create an integrated model of reality consisting of representations of signals from the environment and movement in this environment. These representations are wave patterns in which elements distributed over the neural network participate. If the physical nature of all representations is the same, then it is logical that the technology of their creation, storage and reproduction is the same, despite the participation of various elements of the network. Moreover, physiologically, the neural populations of the cerebral cortex involved in forming different representations are similar, and the differences are not fundamental.

This physical, physiological and technological unity has several advantages. First, the wave nature of the process allows representations to be dynamic structures, simultaneously associated with the elements that create them, but also free from them in the sense of the possibility of changing localization. Second, the homogeneity of the substrate allows the components to be interchangeable. Third, neurons and their populations are filters with specific impulse characteristics, and their plasticity is just a retuning and is not an insoluble technological problem, although it requires time and energy.

In general, TTT considers the cortex as an integrating filter and its functional separation as different levels of integration: primary, intermediate, and higher. From this point of view, it is not surprising that, if necessary, integrating filters can process information from different sources. The task of transducers in sensor systems is the primary processing of signals. If they have coped with it, then the universal processor of the cortex can dynamically reconfigure to create representations from the incoming code, regardless of what was the original signal.

Some animals have sensory systems that operate in a range beyond our reach — for example, ultrasonic echolocation, electroreceptors, or night vision. Theoretically, by creating an appropriate interface of artificial systems for processing such signals with the brain, we can acquire "super senses" since the capabilities of our cortex integrators are very high.

We are already creating substitutes for our natural sensory systems in the event of damage and pathology. For example, the development of cochlear implants has been going on since the late 1950s. The first devices had only one electrode, which generated signals from the sound wave and transmitted them to the receptors of the auditory nerve. But one channel was not enough to create at least the minimum frequency range of sounds. This device now consists of two parts: a microphone, microprocessor and transmitter that process sound waves, and a receiver with an array of electrodes installed subcutaneously and connected to the cochlea through surgery. The first part replaces the outer ear and part of the inner ear as a primary transducer of sound waves into mechanical and hydraulic vibrations. The second part replaces the function of the cochlea to create electrical impulses from these oscillations for further encoding by the brain in representation. Of course, even modern devices with dozens of electrodes cannot compare with natural technology with its thousands of electrodes (cochlear hair receptors). The result leaves much to be desired, but it returns at least part of the world's music to the deaf.

We are accustomed to the fact that there are ways to improve the visual pathway's optical part. Glasses appeared several centuries ago. In the twentieth century, we began to perform operations to enhance the state of the optics of the eye. Without this, a vast number of people would simply go blind. But this is only the beginning of the journey. We are already moving deeper into the chain. In 2015, the first bionic eye implantation was performed. The surgeons implanted electrodes in the eye retina of a patient with age-related degenerative visual impairment. They connected them to a miniature video camera with wireless data transmission installed into glasses. This is no longer just an improvement but a replacement for the natural signal converter. We become human cyborgs thanks to the technological approach to our brains.

The first legally recognized human-cyborg was Neil Harbisson, who implanted an antenna in his head in 2004. Bioethics committees repeatedly rejected the operation, but anonymous doctors did it. The antenna allows him to feel and hear colors as sound vibrations inside the head, including infrared and ultraviolet radiation invisible to the human eye. The antenna also provides Internet connectivity and hence color reception from distant sources. In 2010, Harbisson co-founded the Cyborg Foundation, an international organization that defends the rights of cyborgs, promotes the art of cyborgs, and supports people who want to become cyborgs.

Harbisson's antenna is more of an ethical than a technological breakthrough since it acts simply as a primary converter of one type of signal to another. We all hear sounds not only with our ears but also with our heads. The skull acts as a resonator, and the ear perceives its vibrations. That is why the sound of our voice, in which usually the vibrational frequencies of the cranium play a large part, seems to us completely different if we hear it in the recording since with external perception, the frequency input of the skull is much less.

To perceive the vibrations created by the antenna as a color, the brain only needs training in creating visual representations from sounds. Harbisson, who has an achromatic vision from birth due to the pathology of eye sensors, required a long adaptation. He first learned to perceive seven notes as seven colors, then gradually, more subtle nuances came. By the way, some people can perceive colors as sounds or vice versa, and in principle, have a mixed picture of the world. This phenomenon is called synesthesia (from the ancient Greek σύν "together," and αἴσθησις "sensation").

It is usually defined as "a perceptual phenomenon in which stimulation of one sensory or cognitive pathway leads to involuntary experiences in a second sensory or cognitive pathway" (Wikipedia "Synesthesia"). There are different variants: people see colors when they hear sounds; perceive odors when they listen to sounds; hear sounds when they see movement; feel tastes when listening, etc.

The above definition may be a bit misleading. The phenomenon does not arise at the level of senses but at the integration level in the cortex. On the one hand, it shows the universal nature of this processor again. On the other hand, in itself, involuntary mixing means a disruption in the operation of the system since we usually perceive the world, albeit as a whole, but still differentiated by modalities.

We can, of course, turn on our imagination and see colors or even images while listening to music. We can associate numbers with color and count by drawing multi-colored pictures. Mnemonics are all based on associative thinking. But the fundamental feature of genuine synesthesia is its involuntary nature. Synesthete will see color music even if he doesn't want to.

Pronounced synesthesia is a relatively rare condition (less than 1% of people). In any case, in principle, this is a violation of normal functioning. Why? Is it so bad to see color music? It is interesting in this sense to look at the attitude of the carriers of the phenomenon themselves. Those who were born and grew up as synesthetes sometimes simply do not know that their vision of the world is different. When they finally realize that, they still perceive it as normal. They just don't know the other. They cannot imagine what it is like not to see sounds. For them, the question "How do you feel about it?" is strange, meaningless. But those who have acquired this during their lifetime due to some violation in the functioning of the substrate perceive synesthesia as a very disturbing acquisition.

So, why such a state is maladjustment? Maybe such a mixed-up world is better? The question is not whether such a world is better or worse. The question is the adaptability of such a mixed reality model. Let's put the question differently: why aren't most humans synesthetes? Why is our perception differentiated by modality, even though there is a general picture of the world?

Imagine that a hunter walks through the forest in search of prey. This is almost a fantastic picture for many modern people, but it is quite relevant for other living beings. So, he walks and hears a sound. But together with it and with the general visual picture of the forest, "color music" appears. If there are many sounds, then everything will look like a disco hall, where colored lights flicker. What if these sounds belong to potential prey? What are the chances of catching it if there are disco lights in front of the hunters' eyes? The chances of getting a hare for dinner are falling rapidly. And if it's not a hare, but someone bigger? Then, on the contrary, the chances of the hunter becoming prey increase. From whichever side you look, this is a maladaptive situation. And what if real sounds mixed with phantom smells? Not only do they knock off the trail of the actual scent, but they also create false targets.

But let's go back to the city and get behind the wheel of the car. Environmental sounds give us clues about what is happening, often even earlier than visual cues. Imagine that these sounds evoke "color music" or other images that have nothing to do with the real sources of the sounds themselves. The chances of getting from point A to point B are falling rapidly.

Synesthesia is a variant of the maladaptive state of the system. The majority (statistical norm) is not synesthetes. But initially, at birth, we are all likely synesthetes. The primary converters have not yet fully tuned in to the signals, and the subsequent filters have not yet received their "job." Of course, a lot is defined genetically. Still, at the initial stage of development of the nervous system, the very rapid growth of connections takes place, and only then are the zones tuned to specific signal ranges. Gradually, feelings acquire their specialization; the world becomes more differentiated. Typically, everything diverges along its intended channels and then falls into higher-level integrators, which bring all the ranges together but retain the polyphonic depth. The synchronized model of reality is not a unison-mixing but a structure filled with harmonies, in which each party has its own identity.

To understand the world of the synesthete, one can take the case of Solomon Shereshevsky, described by Alexander Luria in the book "The Mind of a Mnemonist: A Little Book about a Vast Memory)" (Luria, 1968). Shereshevsky demonstrated astounding memorization capabilities. But this did not make his intelligence outstanding and, in many ways, pathologically influenced his cognitive abilities. He could not memorize information according to its meaning. The images that helped him to show the wonders of mnemonics often did not allow him to associate information with its true meaning. Mnemonics were available to him, but semantics were not.

He had problems with distinguishing faces since they were full of changeable sensations for him. He could hardly read because the letters could evoke visions entirely unrelated to the text. Imagine that letters paint words in colors or create sounds, but not of words in the book, but of a completely different content, or evoke a taste sensation. It was especially difficult to eat and read at the same time, as the result was an unimaginable mess of taste. Speaking of taste, when he once wanted to buy ice cream, and the saleswoman offered him a fruit flavored one, the sound of her voice caused him to see black pieces of coal falling out of her mouth. After that, naturally, he didn't want to eat ice cream.

In general, synesthesia can hardly be called an adaptive state, and multiple synesthesia, like Shereshevsky's, seriously complicates life. Only the support of society allows such people to survive. The synesthete cheetah would hardly have survived to an independent age if antelopes for it turned into black coal. The interchangeability of the integrators has advantages when it is necessary to compensate for the lost function, but normally in this orchestra, each musician should play a specific part. The music of the Mind must be differentiated and integrated at the same time.

But let us return to the combination of artificial signal transduction technologies with natural ones. In this process, we are still at the level of primary converters. Is there a problem with extending this approach to the further chain? Of course, there are substantial technical difficulties, but there are more conceptual difficulties without resolving which technological progress is impossible. The fact that external sensors work with signals from the environment and convert them into impulses has been known for so long that in this sense, there is no theoretical obstacle for artificial technologies.

But further it is much more difficult because, in principle, there is no concept of what is happening along the entire internal technological chain. The fact that these impulses, as a result of the work of the subcortical and cortical structures, turn out to be representations of sound has also been understood for a long time. The problem is in understanding and explaining the function that these brain structures perform. Moreover, the problem is in the definition of the function itself.

With all the technical complexity of the visual, auditory and other implants described above, they are suitable only for pathologies at the first signal processing stages. They are not directly related to the brain code but simply provide the brain with "food for thought." It only needs an adaptation period to correlate the incoming signals with the existing model, as a matrix projected onto them, and correct this model if necessary. It is no coincidence that all patients take time to see or hear again.

But if the pathology is at the stages of the encoding itself, then we cannot do without understanding the technologies and algorithms of the Mind. Technical solutions will follow conceptual ones, and then we are faced with a real prospect of treating such disorders. We are talking not only about the sensory-motor part but also about systemic disorders, which are usually called mental illnesses.

The problem in the treatment of pathologies of the Mind is more conceptual than technical. We are accustomed to the fact that operations on the implantation of artificial organs have become almost standard: mechanical (for example, joints), hydraulic (for example, organs of the cardiovascular system), and others. However, everything related to the brain so far borders on "miracles."

Many neuroscientists admit that it is much easier for them to imagine replacing a joint or a heart valve with a prosthesis than a similar operation on the brain. The problem is not technical but conceptual. They say: "I can believe in such a replacement for the knee, but the question arises about the brain. Under this assumption lies the hypothesis that the brain is an information processing device, and its elements can be replaced to perform the same function. There is no certainty that this hypothesis is correct." This anonymous quote is from a speech of the chairman of one of the international associations in the field of neuroscience at a congress in 2016.

But concerning the knee or the heart, one does not have to believe, since this is already a fact of our life. We know their function, which means we can model the technology. A hundred or two hundred years ago, this required a leap of faith. All that is necessary now is the patient's consent to the operation. What's the problem with the brain issue? The question is about the function: what for?

Suppose it is about processing signals and creating information from them in order to carry out actions based on this information. In that case, everything becomes more or less clear, and we can answer further questions, which boil down to a single technological one: how?

For some, this is an obvious hypothesis, but for others, it is very controversial. But if there is no certainty that such an assumption is correct, then a counter-question arises: what is the brain for, if not for this? There have been many answers over the millennia, but for the most part, they were far from the realities of physics and technology. If we do not think of technology as a way of solving the problems facing the brain, then all we have to do is repeat the phrase "poorly understood" in every way.

Let's take a simple example in which the brain's technological chain is manifested. Let's go back 200 years when a little boy named Louis Braille lost his sight and came up with a way to read using the embossed paper method. It was based on "night writing" methods used by the military to read reports in the dark. Now we are so accustomed to this way of reading by the blind that we are not even surprised by this example of brain plasticity and of the fact that the brain's task is precisely to process the signals from which it creates information. After all, what an ordinary person sees with his eyes, a blind person sees with his hands. But in reality, both see by the brain, a device for processing signals and creating information from them.

A blind person uses a stick as an extension of a hand for feeling objects and for echolocation. This is seeing things through ears. Even some non-blind people have developed this ability to such an extent that they can see the outline of an object with their eyes closed. It's not the mystical "third eye," but the Mind's eye: integrating filters as creators of representations. Any sonic echolocation reflections provide enough information for such a person to create a rough but working model of the object. You can flick your finger, tongue or simply pick up sounds from other external sources to use the reflected sound waves to create a visual representation. And there is no mystery about it if we approach the work of the brain technologically.

Paul Bach-y-Rita, a brilliant engineer and an equally brilliant neuroscientist who devoted his work to the dynamics and plasticity of the brain, admitted at the end of his life: "The mechanisms of representational changes are not known ... Although sensory substitution studies strongly support the capacity of the brain to reorganize, the actual mechanisms have not been firmly established" (Bach-y-Rita, Kercel, 2003).

Here is how Nikolai Bernstein described the dilemma: "Nerve impulses in afferent fibers, easily recorded with the help of modern technology, do not contain anything similar either to light, or heat, or a mechanical force of tension (acting on proprioceptors), but only certain sequences of peaks of bioelectric action potentials, as far as we judge with today's level of experimental technology ... Meanwhile, it is clear that it is these chains of impulses that provide the perception of all the qualities of the surrounding world — light, colors, sound tones, and the entire wealth of skin, olfactory, gustatory, kinesthetic sensations. It is obvious that this richest sensory information about the external world is communicated to our brain encoded in the form of impulse chains. How the brain decodes it, is there a specific difference in nerve signals from visual, auditory, tactile and other receptors and what it consists of — these are tasks that the physiology of tomorrow has yet to solve" (Bernstein, 1966).

But this requires overcoming the dead-ends of the old paradigms. "Sometimes the reason for the urgent need to change the theory is the gradual accumulation of data that does not fit into its framework; sometimes it is a single fact or phenomenon that strikes it to the very heart. It also happens that science has to wait a long time for the moment until a brilliant seer appears on the battlefield, who will be able to find and formulate a fresh, powerful concept that kills previous views by the mere fact of its indisputable advantage over them ... In other cases, a change in natural science theories takes place not so revolutionary. In this category of cases, the replacement of the old theory does not occur because its erroneousness or illegitimacy has been revealed. The young concept that comes to replace it takes over, either because it is able to generalize a much wider range of phenomena than before or because it turns out to have a much greater heuristic power" (Ibid).

Teleological Transduction Theory is able to generalize a wide range of phenomena and explain them down to subtle details. Using the musical analogy, we can say that it plays a symphony that has a wide and deep polyphony and polyrhythm. It tries to be up to the complex symphony of the Mind that is played by the brain.

In this and the previous volume of the "Symphony of Matter and Mind" series, we answered many questions, including the one that Bach-y-Rita and Bernstein asked: the way of encoding and decoding the qualities of the surrounding world; a mechanism for creating and changing representations. An important question remains: how do representations of different signals combine into a single model of reality, while keeping their identity intact? In the language of musical analogy: how do different notes merge into a single symphony while maintaining their individual sound? This will be the theme of the next volume.

ANNEXES

Visual System Lemniscate

Processes:

1. The effect of light waves reflected from objects on receptors.
2. Signal processing — transduction, transmission, modulation, synchronization, creation of representations, storage, reproduction.

Perception mechanisms (primary processing and transformation of signals):

1. Photoreceptors (rods, cones).
2. Bipolar cells.
3. cGMP (Cyclic guanosine monophosphate) ion channels Ca^{++}, Na^+, K^+.
4. Conformational change: opsin/rhodopsin in cGMP.
5. Activation/deactivation of the ion channel.
6. Ganglion cells "on-center" and "off-center."
7. Receptive fields — the spatial structure of receptor organization.
8. Horizontal cells as filters-transducers in receptive fields.

Lemniscate stages:

1. Deductive projection of the reality model. Visual cortical structures such as extrastriate visual cortex (zones V3, V4, V5, MT), Parietal Lobe (spatial analysis — Where?), inferior temporal lobe (identification and characteristics —

What?), occipitotemporal gyrus (identification — Who?), and in the frontal cortex create and project a reality model.

2. Inductive projection: the reference waves of the reality model interact with the object waves of the introjected signals.

3. Inductive introjection of environmental signals. Specialized receptors perceive signals (light waves), transform them into an internal code, as patterns of neural activity, and transmit the created patterns to the corresponding thalamus module (lateral geniculate body) as a distribution relay and filter-modulator.

4. Abductive introjection. The information is further distributed among the layers and zones of the cortex for processing, integrating, evaluating, comparing, saving, reproducing and correcting the states of the system and the reality model, transmission of signals to motor zones and effectors for action in accordance with the model.

Auditory System Lemniscate

Processes:

1. The impact of sound waves on the conversion channels in the outer and middle ear.
2. Conversion of a sound signal into a mechanical one.
3. Conversion of a mechanical signal into an electrochemical one.
4. Transmission, modulation, synchronization, integration, storage, reproduction.

Perception mechanisms:

1. Outer ear — focusing the signal.
2. Middle ear — biomechanical amplification and signal transmission.
3. Inner ear — decomposition of sound into frequency components (tonotopy), the transformation of a mechanical signal into an electrochemical one (mechanically gated Na^+ ion channels in stereocilia of hair cells).

Lemniscate stages:

1. Deductive projection of the reality model. Representations in specialized cortical structures (belt of auditory areas, primary auditory cortex, secondary auditory cortex) and the cortex's higher associative zones create and project a model of reality.
2. Inductive projection of the reality model. Information streams are distributed over the nuclei of the thalamus (medial geniculate nucleus) and transmitted to the midbrain (inferior colliculus) and brainstem (pons nucleus of lateral lemniscus, superior olivary complex, rostral medulla cochlea nuclei). Result: the reference waves of the reality model enter into interaction with the object waves of the introjected signals.

3. Inductive introjection of environmental signals. Specialized receptors perceive changes in the environment; the signal is distributed according to the tonotopic principle into frequencies, neurons in spiral ganglion cells transmit along cranial nerve VIII and further to the rostral medulla, pons, midbrain, thalamus. The signals reach the primary auditory cortex and are converted into an acoustic picture of the world.

4. Abductive introjection. The signal is further distributed among the layers and zones of the cortex for processing, integration of synchronization, transmission, evaluation, comparison, preservation, reproduction, correction of the state of the system and the reality model, transmission of signals to motor zones and effectors for action in accordance with the model.

Olfactory System Lemniscate

Processes:

1. The effect of waves of chemicals in the air on the receptors of the epithelium.
2. Conversion of a chemical signal into an electrochemical one.
3. Transmission, modulation, synchronization, integration, storage, reproduction.

Perception mechanisms:

1. Hair receptors with molecular-specific receptors in the membrane (olfactory cilia).
2. Channels with chemical modulation parameters (ligand-gated ion channels).
3. Voltage-gated ion channels.

Lemniscate stages:

1. Deductive projection of the reality model. Representations in the olfactory cortical structures (olfactory cortex, piriform cortex) and the cortex's higher associative zones create and project a model of reality.
2. Inductive projection of the reality model. Streams of information come to the olfactory bulb (structure, concentration, frequency response). Result: the reference waves of the reality model enter into interaction with the object waves of the introjected signals.
3. Inductive introjection of environmental signals. Specialized receptors perceive environmental changes (biochemical transformation, conformational change in proteins), release (exocytosis) of neurotransmitters, activation of SMS, the opening of Ca^{++}, Na^+ channels, the influx of cations, depolarization, the opening of Cl-channels, outflow of anions, release of Ca^{++} from intracellular

stores, increased depolarization, action potential, signal transmission. The signals reach the olfactory bulb and are converted into an olfactory picture of the world.

4. Abductive introjection. The signal is further distributed among the layers and zones of the cortex for processing, integration of synchronization, transmission, evaluation, comparison, preservation, reproduction, correction of the state of the system and the reality model, transmission of signals to motor zones and effectors for action in accordance with the model.

Gustatory System Lemniscate

Processes:

1. The effect of waves of chemicals in solution on epithelial receptors.
2. Conversion of a chemical signal into an electrochemical one.
3. Transmission, modulation, synchronization, integration, storage, playback.

Perception mechanisms:

1. Molecular-specific receptors (taste cells).
2. Channels with chemical modulation parameters (ligand-gated ion channels).
3. Voltage-gated ion channels.

Lemniscate stages:

1. Deductive projection of the reality model. Representations in taste cortical structures (insular taste cortex, frontal taste cortex) and the higher associative zones create and project a model of reality.
2. Inductive projection of the reality model. Streams of information are distributed over the nuclei of the thalamus and come to the brainstem, nucleus of the solitary tract (amplitude-frequency characteristics of signals, spatial topography of taste buds). Result: the reference waves of the reality model enter into interaction with the object waves of the introjected signals.
3. Inductive introjection of environmental signals. Specialized receptors (salinity, acidity, sweetness, bitterness, umami) perceive environmental changes (biochemical transformation, conformational change in proteins), release (exocytosis) of neurotransmitters, activation of SMS, the opening of Ca^{++}, Na^+ channels, the influx of cations, depolarization, the release of Ca^{++} from intracellular stores, increased depolarization, action potential, signal transmission.

Topographic organization of receptors, labeled lines, filters, and thresholds of perception by concentration at initial introjection and transformation level.

4. Abductive introjection. The signal is further distributed among the layers and zones of the cortex for processing, integration of synchronization, transmission, evaluation, comparison, preservation, reproduction, correction of the state of the system and the reality model, transmission of signals to motor zones and effectors for action in accordance with the model.

Mechano-Sensory Lemniscate
Dorsal Column Medial Lemniscus Pathway

Processes:

1. Mechanical effect on receptors.
2. Conversion of a mechanical signal into an electrochemical one.
3. Signal processing — modulation, amplification, transmission, synchronization, storage, reproduction.

Perception mechanisms:

1. Receptors — Meissner corpuscle, Merkel cells (touch), Ruffini corpuscle (stretch), Pacinian corpuscle (vibration).
2. Stretch-gated ion channels.
3. Receptive fields.
4. Labeled lines (signal transmission by parallel channels).

Lemniscate stages:

1. Deductive projection of the reality model. Somatotopy in the primary somatosensory cortex (postcentral gyrus area) defines the body map, the centers for storing and reproducing information determine the representations of the map of the body and space, the centers of emotional systems determine the valence of representations. Somatotopy at the level of the modulators (somatosensory areas of thalamus, ventral posterior lateral nucleus of thalamus) distributes it along parallel lines to the brainstem (caudal medulla area).
2. Inductive projection of the reality model from brainstem with further distribution along parallel labeled lines to receptors. Result: the reference waves of the reality model enter into interaction with the object waves of the introjected signals.

3. Inductive introjection of environmental signals. Specialized receptors and ion channels perceive changes in the environment, the first-level neurons in dorsal root ganglion cells are transmitted along a chain in the cuneate or gracile tract (depending on the level of entry into the spinal tract) to the bifurcation point in caudal medulla (cunate nucleus, gracile nucleus) and second-level neurons.

4. Abductive introjection. The signal reaches the third level neurons in the VPL (thalamus), where it is distributed along the lines to the PSSC, enters the fourth layer of the cortex, and is further distributed among the layers and zones of the cortex for processing, integration of synchronization, transmission, evaluation, comparison, preservation, reproduction, adjusting the state of the system and the model of reality, transmitting signals to motor zones and effectors to act in accordance with the model.

Pain System Lemniscate
Anterolateral System of Sensorimotor Pathways

Processes:

1. Transformation of mechanical, chemical, thermal signals into electrochemical ones.
2. Signal processing — modulation, amplification, transmission, synchronization, storage, reproduction.

Perception mechanisms:

1. Free nerve endings as polymodally gated transient receptor potential ion channels. Ca^{++}, Na^+ channels are signal transducers of thermal, light, chemical, mechanical origin.
2. Receptive fields.
3. Labeled lines.

Lemniscate stages:

1. Deductive projection of the model of reality and the state of the system. General representations in the associative areas of the cortex (higher integrators) and detailed representations of the body in the primary somatosensory cortex, post central gyrus area. Somatotopy at the level of filters-modulators: somatosensory areas of thalamus, ventral posterior lateral nucleus of the thalamus, distributes it along parallel lines to the brainstem.
2. Inductive projection of the reality model. Result: the reference waves of the reality model enter into interaction with the object waves of the introjected and encoded environment signals.
3. Inductive introjection of environmental signals. Polymodal receptors and their ion channels perceive changes in the environment; the signal is converted into patterns of neuronal activity, the first level neurons in the dorsal root ganglion

cells are transmitted along a chain in the dorsal horn to the second level neurons up to the bifurcation point in the brainstem.

4. Abductive introjection. From the bifurcation point, the signal reaches the third-level neurons along the spinothalamic tract into the somatosensory zones of the thalamus, where it is distributed along the lines to the primary somatosensory cortex, enters the fourth layer of the cortex and is further distributed over the layers and zones of the cortex for processing, synchronization integration, transmission, evaluation, comparison, preservation, reproduction, correction of the state of the system and the model of reality, the transmission of signals to motor zones and effectors for action in accordance with the model.

REFERENCES

Bach-y-Rita, P. (2003). *In press: Can You See With Your Tongue?* By Michael Abrams, Dan Winters Sunday, June 01, 2003 DiscoverMagazine.com 02/05/15 15:04 .

Bach-y-Rita, Paul, Kercel, Stephen W. (2003). *Sensory substitution and the human–machine interface.* Trends in Cognitive Sciences Vol.7 No.12 December 2003.

Barbour, J. (1999). *The End of Time: The Next Revolution in our Understanding of the Universe.* Oxford Univ. Press.

Bernstein, N. (1962). *New lines of development in physiology and their relationship with cybernetics.* Institute of Philosophy of the Academy of Sciences of the USSR. Moscow.

Bernstein, N. (1990). P*hysiology of movements and activity.* Reprint of monographs 1947, 1966 Moscow, Science.

Borelli, D. (1680). *On the movement of animals (De Motu Animalium).* Rome, 1680

Buzsáki, Gyorgy, Peyrache, Adrien, Kubie, John (2014). *Emergence of Cognition from Action.* Cold Spring Harbor Laboratory Press doi: 10.1101/sqb.2014.79.024679. Cold Spring Harbor Symposia on Quantitative Biology, Volume LXXIX 1.

Cajal, S. (1911). *Histologie du Syste ́me Nerveux de l'Homme et des Verte ́bre ́s.* Paris: Maloine.

Campbell, F.W. Robson, J.G. (1968). *Application of Fourier analysis to the visibility of gratings.* J.Physiol. 1968, 197, pp. 551-566.

Chalmers, D. J. (1995). *Facing up to the Problem of Consciousness.* Journal of Consciousness Studies 1995 Vol.2, № 3 P. 200-219.

Cheron, G., M. Duvinage, C. De Saedeleer, T. Castermans, A. Bengoetxea, M. Petieau, K. Seetharaman, T. Hoellinger, B. Dan, T. Dutoit, F. Sylos Labini, F. Lacquaniti, Y. Ivanenko. (2012). *From Spinal Central Pattern Generators to Cortical Network: Integrated BCI for Walking Rehabilitation.* Hindawi Publishing Corporation Neural Plasticity Volume 2012, Article ID 375148.

Deadwyler SA, Hampson E, Forest W, Carolina N, Bunn T, Hampson RE. (1996) *Hippocampal ensemble activity during spatial performance in rats.* J Neurosci 16:354-372.

Deitch D, Rubin A, Ziv Y. (2021). *Representational drift in the mouse visual cortex.* Current Biology. 2021

Descartes, R. (1637-1647). *Works in 2 volumes.* Moscow. Beginnings of Philosophy 1989-1984.

De Valois, R., Albrecht, D. G., Thorell, L. G. (1982). *Spatial frequency selectivity of cells in macaque visual cortex.* Vision Res. Vol. 22, pp 545-559.

De Valois, R. L., De Valois, K. K. (1988). *Spatial Vision.* New York: Oxford University Press.

Dhawale Ashesh K, Poddar Rajesh, Wolff Steffen BE, Normand Valentin A, Kopelowitz Evi, Ölveczky Bence P. (2017). *Automated long-term recording and analysis of neural activity in behaving animals.* eLife, 6:e27702, 2017.

Driscoll LN, Pettit NL, Minderer M, Chettih SN, Harvey CD. (2017). *Dynamic reorganization of neuronal activity patterns in parietal cortex.* Cell. 2017; 170(5):986–999.

Engvig A. et al. (2012). *Memory training impacts short-term changes in aging white matter: a longitudinal diffusion tensor imaging study.* Human Brain Mapping 33, 2390-2406.

Fairhall, Adrienne L., Burlingame, C. Andrew, Narasimhan, Ramesh, Harris, Robert, Puchalla, Jason L., Berry, Michael J. (2006). *Selectivity for Multiple Stimulus Features in Retinal Ganglion Cells.* J Neurophysiol 96: 2724–2738.

Fodor, J., Pylyshyn, Z. (1998) *Connectionism and cognitive architecture: A critical analysis.* Cognition, Volume 28, Issues 1–2, 1988, Pages 3-71.

Ford MC, Alexandrova O, Cossell L, Stange-Marten A, Sinclair J, Kopp-Scheinpflug C, Pecka M, Attwell D, Grothe B. (2015). *Tuning of Ranvier node and internode properties in myelinated axons to adjust action potential timing.* Nat Commun. 2015 Aug 25;6:8073.

Freeman, W. J. (1992). *Tutorial on neurobiology: from single neurons to brain chaos.* International Journal of Bifurcation and Chaos 2: 451-482. 1992

Freeman, W. J. (2008). *Nonlinear Brain Dynamics and Intention According to Aquinas.* Mind & Matter Vol. 6(2), pp. 207–234.

Freusberg, A. (1874). *Reflexbewegungen beim Hunde.* Pflueger's Archiv fuer die gesamte Physiologie, 9: 358–391.

Fujisawa, Shigeyoshi, Buzsáki, Gyorgy (2011). *A 4 Hz Oscillation Adaptively Synchronizes Prefrontal, VTA, and Hippocampal Activities .* Neuron 72, 153–165, October 6, 2011.

Fukuma, R, Yanagisawa, T, Yorifuji, S, Kato, R, Yokoi, H, Hirata, M, et al. (2015). *Closed-Loop Control of a Neuroprosthetic Hand by Magnetoencephalographic Signals.* PLoS ONE 10(7): e0131547.

Gabor, D. (1948). *A new microscopic principle.* Nature, 1948, V.161, pp.777-778

Graziano, M.S.A., Taylor, C.S.R. and Moore, T. (2002). *Complex movements evoked by microstimulation of precentral cortex.* Neuron. 34 (5): 841–851

Griffin, D.M., Hoffman, D.S., Strick, P.L. (2015). *Corticomotoneuronal cells are «functionally tuned».* Science.2015 Nov 6;350(6261):667-70. doi: 10.1126/science.aaa8035.

Grillner, S, Zangger, P. (1975). *How detailed is the central pattern generation for locomotion?* . Brain Res 1975; 88:367–71. PubMed: 1148835.

Grillner, S., Wallen, P., Brodin, L., Lansner, A. (1991). *Neuronal network generating locomotor behavior in lamprey: circuitry, transmitters, membrane properties, and simulation.* A. Rev. Neurosci. 14, 1699199.

Heitmann, Stewart, Boonstra, Tjeerd, Breakspear, Michael (2013). *A Dendritic Mechanism for Decoding Traveling Waves: Principles and Applications to Motor Cortex.* PLOS Computational Biology Volume 9. Issue 10. October 2013.

Hochberg, L. D. (2012). *Reach and grasp by people with tetraplegia using a neurally controlled robotic arm.* Nature. 2012 May 16;485(7398):372-5. doi: 1.

Hofstetter, S., Tavor, I., Moryosef, S.T., Assaf, Y. (2013). *Short-term learning induces white matter plasticity in the fornix.* J. Neuroscience 33, 12844-50.

Hubel, D. H., Wiesel, T. N. (1959). *Receptive fields of single neurones in the cat's striate cortex* . The Journal of Physiology. 124 (3): 574–591.

Hubel, D. H. (1988). *Eye, Brain, and Vision.* Scientific American Library, distributed by W. H. Freeman & Co., New York, 1988

Huber Daniel, Gutnisky Diego A, Peron Simon, O'connor Daniel H, Wiegert J Simon, Tian Lin, Oertner Thomas G, Looger Loren L, Svoboda Karel. (2012). *Multiple dynamic representations in the motor cortex during sensorimotor learning.* Nature, 484(7395):473, 2012.

Ito, R; Lee, AC (2016). *The role of the hippocampus in approach-avoidance conflict decision-making: Evidence from rodent and human studies.* Behavioural Brain Research 15 October 2016, 313: 345–57. doi:10.1016/j.bbr.2016.07.039. PMID 27457133.

Jackson, F. (1982). *Epiphenomenal Qualia.* Philosophical Quarterly. 32: 127–136.

Jackson, Andrew, Fetz, Eberhard E. (2012). *Interfacing with the Computational Brain.* NIH Public Access Author Manuscript IEEE Trans Neural Syst Rehabil Eng. Author manuscript; available in PMC 2012 June 11.

Jankowska, E, Jukes, MGM, Lund, S, Lundberg, A. (1967). *The effect of DOPA on the spinal cord 5. Reciprocal organization of pathways transmitting excitatory action to alpha motoneurones of flexors and extensors.* Acta Physiol Scand 1967; 70:369–388. PubMed: 429347.

Katlowitz Kalman A, Picardo Michel A, and Long Michael A. (2018). *Stable sequential activity underlying the maintenance of a precisely executed skilled behavior.* Neuron, 2018.

Kim, Yoon Jae, Park, Sung Woo, Yeom, Hong Gi, Bang, Moon Suk, Kim, June Sic Chun, Chung, Kee, Kim, Sungwan (2015). *A study on a robot arm driven by three-dimensional trajectories predicted from non-invasive neural signals.* BioMed Eng OnLine 14:81.

Korzybski, Alfred (1933). *Science and Sanity: An Introduction to Non-Aristotelian Systems and General Semantics.* International Non-Aristotelian Library Publishing Company.

Kupers, R. et al. (2003). *Activation of visual cortex by electro-tactile stimulation of the tongue in early-blind subjects* . Neuroimage 19, S65.

Leith, E.N., Upatnieks, J. (1964). *Wavefront reconstruction with diffused illumination and three-dimensional objects.* J. of the Optical Society of America, 1964, V. 54, p.1295.

Levine, J. (1983). *Materialism and Qualia: the Explanatory Gap.* Pacific Philosophical Quarterly. 1983. Vol. 64, № 4. P. 354—361.

Lewis, Clarence Irving (1929). *Mind and the world-order: Outline of a theory of knowledge.* New York: Charles Scribner's Sons. p. 121

Llinas, R. (2001). *I of the Vortex: From Neurons to Self.* The MIT Press.

London, Michael, Hausser, Michael (2005). *Dendritic Computation.* Annu. Rev. Neurosc. 28:503-32.

Lundberg, A. (1981). *Half-centres revisited.* In: Szentagotheu, J.; Palkovits, M.; Hamori, J., editors. Regulatory Functions of the CNS. Motion and Organization Principles. Budapest, Hungary: 1981. p. 155-167. Pergamon Akademiai Kiado, Adv. Physiol. Sci.

Luria, A.R. (1968). *The Mind of a Mnemonist: A Little Book about a Vast Memory.* Moscow University Publishing House.

Luria, A.R. (1973). *Fundamentals of Neuropsychology.* Moscow.

MacDonald CJ, Lepage KQ, Eden UT, Eichenbaum H. (2011). *Hippocampal "time cells" bridge the gap in memory for discontiguous events.* Neuron 71:737-749.

Mahan, Margaret Y., Georgopoulos, Apostolos P. (2013). *Motor directional tuning across brain areas: directional resonance and the role of inhibition for directional accuracy.* Front. Neural Circuits, 15 May 2013.

Marks TD, Goard MJ. (2021). *Stimulus-dependent representational drift in primary visual cortex.* Nature communications. 2021; 12(1):1–16.

McCreaa, David A., Rybak, Ilya A. (2008). *Organization of mammalian locomotor rhythm and pattern generation.* Brain Res Rev. 2008 January; 57(1): 134–146.

McCulloch, WS, Pitts, W. (1943). *A logical calculus of the ideas immanent in nervous activity.* The bulletin of mathematical biophysics. 5(4):115–133.

McKenzie, I.A., Ohayon, D., Li, H., Paes de Faria, J., Emery, B., Tohyama, K., Richardson, W. . (2014). *Motor skill learning requires active central myelination.* Science 346. 318-22. 10.1126/science.1254960.

Minsky, M. (1998). *Edge interview with Marvin Minsky.* Edge.org. 1998-02-26.

Moore, Jason J., Ravassard, Pascal M., Ho, David, Acharya, Lavanya, Kees, Ashley L., Vuong, Cliff, Mehta, Mayank R. (2017). *Dynamics of cortical dendritic membrane potential and spikes in freely behaving rats.* Science 24 Mar 2017: Vol.355, Issue6331, eaaj1497

Moravec, H. (1991). *The Universal Robot.* July 1991, Robotics Institute Carnegie Mellon University Pittsburgh, PA 15213 USA www.frc.ri.cmu.edu/~hpm/project.archive/robot.papers/1991/Universal.Robot.910618.html.

Moritz, Chet T., Perlmutter, Steve I., Fetz, Eberhard E. (2008). *Direct control of paralyzed muscles by cortical neurons.* Nature 2008 December 4; 456(7222): 639–642.

Nagel, Thomas (1974). *What is it like to be a bat?* Philosophical Review. LXXXIII (4): 435–450. Oct 1974.

Nakazono T, Sano T, Takahashi S, Sakurai Y. (2015). *Theta oscillation and neuronal activity in rat hippocampus are involved in temporal discrimination of time in seconds.* Front Syst Neurosci 9:1- 12.

Nakazono T, Takahashi S, Sakurai Y. (2019). *Enhanced Theta and High-Gamma Coupling during Late Stage of Rule Switching Task in Rat Hippocampus.* Neuroscience. 2019 Aug 1; 412: 216-232.

Naselaris, Thomas, Prenger, Ryan J., Kay, Kendrick N., Oliver, Michael, Gallant, Jack L. (2009). *Bayesian Reconstruction of Natural Images from Human Brain Activity.* Neuron 63, 902–915, September 24, 2009 Elsevier Inc.

Nirenberg, S., Carcieri, S. M., Jacobs, A. L., Latham, P. E. (2001). *Retinal ganglion cells act largely as independent encoders.* Nature. vol 411 7 june 2001 www.nature.com.

Parker, David, Srivastava, Vipin. (2013). *Dynamic systems approaches and levels of analysis in the nervous system.* Frontiers in Physiology VOLUME 4, 2013.

Pasley, Brian N., David, Stephen V., Mesgarani, Adeen flinker, Shamma, Shihab A., Crone, Nathan E., Knight, Robert T., Chang, Edward F. (2012). *Reconstructing Speech from Human Auditory Cortex.* January 31, 2012 PLOS BIOLOGY on-line http://dx.doi.org/10.1371/.

Pavlov, I.P. (1923). *Twenty years of objective study of the higher activity (behavior) of animals.* Moscow, Science, 1973.

Pribram, Karl (1971). *Languages of the brain; experimental paradoxes and principles in neuropsychology.* Englewood Cliffs, N.J. Prentice Hall

Pribram, K. (1991). *Brain and perception: holonomy and structure in figural processing.* Hillsdale, N. J.: Lawrence Erlbaum Associates.

Pribram, K. (2010). *An instantiation of Eccles brain/mind dualism and beyond.* Complementarity of Mind and Body: Realizing the Dream of Descartes, Einstein and Eccles. 3-10. Nova Science Publishers.

Prochazka, Arthur, Yakovenko, Sergiy (2007). *The neuromechanical tuning hypothesis.* Cisek, Drew & Kalaska (Eds.) Progress in Brain Research, Vol. 165 ISSN 0079-6123.

Purves, D. W. (2010). *Understanding vision in wholly empirical terms.* PNAS 2010, p.6.

Purves, Dale, Augustine, George, Fitzpatrick, David, Hall, William, Lamantia, Anthony-Samuel, White, Leonard (2012). *Neuroscience. 5th Edition.* Sinauer Associates, Inc.: Sunderland, MA. ISBN: (Hardcover) 978-0878936953.

Ramirez, Alexandro D., Ahmadian, Yashar, Schumacher, Joseph, Schneider, David, Woolley, Sarah M. N., Paninski, Liam (2011). *Incorporating Naturalistic Correlation Structure Improves SpectrograReconstruction from Neuronal Activity in the Songbird Auditory Midbrain.* J Neurosci. 2011 March 9; 31(10): 3828–3842.

Read, D. W. (2008). *Working Memory: A Cognitive Limit to Non-Human Primate Recursive Thinking Prior to Hominid Evolution.* Evolutionary Psychology www.epjournal.net 6(4): 676-714 .

Rieke, F., Bialek, W., Warland, D., de Ruyter van Steveninck, R. R. (1999). *Spikes: Exploring the neural code.* MIT Press, Cambridge.

Rubin A, Geva N, Sheintuch L, Ziv Y. (2015). *Hippocampal ensemble dynamics timestamp events in long-term memory.* Elife. 2015; 4:e12247.

Schoonover CE, Ohashi SN, Axel R, Fink AJP. (2021). *Representational drift in primary olfactory cortex.* Nature. 2021 Jun;594(7864):541-546.

Schwartz, Jeffrey, Beyette, Beverly (1997). *Brain Lock: Free Yourself from Obsessive-Compulsive Behavior.* New York: Regan Books, 1997. ISBN 0-06-098711-1.

Sherrington, Charles Scott (1906). *The integrative action of the nervous system* (1st ed.). Oxford University Press: H. Milford.

Shevchenko, V. (2013). *LEONARDO. Second beginning (point of the eye).* 14/10/2013 http://www.topos.ru/article/ontologicheskie-progulki/leonardo-vtoroe-nachalo-tochka-glaza).

Stratton, G. M. (1896). *Some preliminary experiments on vision without inversion of the retinal image.* Psychological Review 3: 611-617.

Stratton G. M. (1897). *Vision without inversion of the retinal image.* Psychological Review, 1897, 4: 341-360, 463-481.

Strogatz, S.H., Stewart, I. (1993). *Coupled oscillators and biological synchronization.* Scientific American 269 (6), December, 68-75. 1993.

Velliste, M, Perel, S, Spalding, MC, Whitford, AS, Schwartz, AB. (2008). *Cortical control of a prosthetic arm for self-feeding.* Nature 2008 Jun 19;453(7198):1098-101. doi: 10.1038/nature06996. Epub 2008 May 28.

von Helmholtz, H. (1867). *Handbuch der physiologischen Optik.* Leipzig, Leopold Voss, 1867 translated as Treatise on Physiological Optics, 1925 by Optical Society of America.

Voronkov, G. (2009). *The intersections of the optic fibers and the inversion of the retina are the neuroanatomical mechanisms of mirror transformations in the visual system.* Proceedings of the XV International Conference on Neurocybernetics, September 23-25, 2009, p. 67.

Wang, A. L.-C. (n.d.). *An Industrial-Strength Audio Search Algorithm.* avery@shazamteam.com Shazam Entertainment, Ltd.

Winkowski, DE, Nagode, DA, Donaldson, KJ, Yin, P, Shamma, SA, Fritz, JB, Kanold, PO (2017). *Orbitofrontal cortex neurons respond to sound and activate Primary Auditory Cortex Neurons.* Cereb Cortex. 2017 Jan 8. doi: 10.1093/cercor/bhw409.

Wood ER, Dudchenko PA, Robitsek RJJ, Eichenbaum H. (2000) *Hippocampal neurons encode information about different types of memory episodes occurring in the same location.* Neuron 27:623-633.

Wu, J.-Y. H. (2008). *Propagating Waves of Activity in the Neocortex: What They Are, What They Do*. Neuroscientist. 2008 Oct; 14(5): 487–502.

Zhang S.-J., Ye J., Couey, J. J., Witter, M., Moser, E. I., Moser, M.-B. (2013). *Functional connectivity of the entorhinal-hippocampal space circuit.* Philosophical Transactions of the Royal Society B: Biological Sciences 2013 Vol. 369.

Ziv Y, Burns LD, Cocker ED, Hamel EO, Ghosh KK, Kitch LJ, et al. (2013). *Long-term dynamics of CA1 hippocampal place codes.* Nature neuroscience. 2013; 16(3):264–266.

Yong, Ed (2021). *Neuroscientists have discovered a phenomenon that they can't explain.* https://www.theatlantic.com/science/archive/2021/06/the-brain-isnt-supposed-to-change-this-much/619145/

Books in the Series
Symphony of Matter and Mind

Part one

Music of Matter
Mechanism of Material Structures Formation

Part two

Theory of Energy Harmony
Mechanism of Fundamental Interactions

Part three

Music of Life
Physics and Technology of Living Matter

Part four

Algorithm of the Mind
Teleological Transduction Theory

Part five

Technologies of the Mind
The Brain as a High-Tech Device

Part six

Harmonies of the Mind
Physics and Physiology of Self

Part seven

Inner Universe
The Mind as Reality Modeling Process

Part Eight

Dissonances of the Mind
The Physics of Mental Disorders

About the Author

Stanislav Tregub

Independent researcher.

Research areas: physics, biophysics, neuroscience, psychology, psychiatry.

www.ingramcontent.com/pod-product-compliance
Lightning Source LLC
Chambersburg PA
CBHW082106220526
45472CB00009B/2062